THE EARLY MIOCENE BUFFALO CANYON FLORA
OF WESTERN NEVADA

The Early Miocene Buffalo Canyon Flora of Western Nevada

Daniel I. Axelrod

UNIVERSITY OF CALIFORNIA PRESS
Berkeley • Los Angeles • Oxford

Volume 135
Issue Date: August 1991

UNIVERSITY OF CALIFORNIA PRESS
BERKELEY AND LOS ANGELES, CALIFORNIA

UNIVERSITY OF CALIFORNIA PRESS, LTD.
OXFORD, ENGLAND

Library of Congress Cataloging-in-Publication Data

Axelrod, Daniel I.
    The early Miocene Buffalo Canyon flora of western Nevada /
Daniel I. Axelrod.
            p.   cm. — (University of California publications in
geological sciences; v. 135)
    Includes bibliographical references.
    ISBN 0-520-09766-1 (paper)
    1. Paleobotany—Miocene. 2. Paleobotany—Great Basin.
3. Paleobotany—Nevada.   I. Title.   II. Series.
QE929.A84   1991   91-4757
560'.178—dc20   CIP

Contents

List of Figures, Tables, and Plates, vi
Acknowledgments, vii
Abstract, ix

INTRODUCTION	1
PRESENT PHYSICAL SETTING	2
GEOLOGY	4
Buffalo Canyon Formation, 4	
Physical Setting, 8	
Taphonomy, 9	
COMPOSITION OF THE FLORA	11
Systematic List of Species, 11	
Allied Modern Vegetation, 16	
PALEOECOLOGY	22
Vegetation, 22	
Climate, 24	
Elevation, 25	
AGE	29
REGIONAL RELATIONS	30
SYSTEMATIC DESCRIPTIONS	34

Appendix, 69
Literature Cited, 71
Plates, 77

List of Figures, Tables, and Plates

Figures

1. Map showing location of the Buffalo Canyon and other western floras referred to, x
2. Geologic map of the Buffalo Canyon area, in pocket
3. Temperature in areas where species show relationship to those in the fossil flora, 26
4. Estimated mean monthly temperatures for the Buffalo Canyon flora, 27
5. Principal vegetation regions during the Middle Miocene, 31
6. Late Middle Miocene floras show marked local floristic changes, 32

Tables

1. Stratigraphic units in the Buffalo Canyon area, 5
2. Number of identified specimens in the Buffalo Canyon flora, 14
3. Distribution of modern species allied to Buffalo Canyon taxa, 16
4. Probable growth habit of Buffalo Canyon species, 17

Plates

1.   Fig. 1. The Buffalo Canyon locality in the lower piñon-juniper belt
    Fig. 2. Allied conifer-sclerophyll vegetation in the Klamath Mountains of northwestern California

2.   Fig. 1. Conifer-hardwood forest in the Adirondack Mountains of New York, with species allied to Buffalo Canyon taxa
    Fig. 2. Conifer-hardwood forest in the Porcupine Mountains of Michigan, with species allied to Buffalo Canyon taxa

3-22.   Buffalo Canyon fossils

Acknowledgments

This contribution is part of a long-term research project to describe and interpret Tertiary floras from the Great Basin province and border areas, a project entirely supported by the National Science Foundation, most recently under Grant BSR 86-07243.

The rich Buffalo Canyon flora was brought to my attention by Peggy Wheat of Fallon, Nevada, who is herewith thanked for her interest and cooperation. Thanks are extended to Katherine J. Barrows and Allan Barrows, who assisted in collecting the flora. Harold F. Bonham helped collect adequate rock samples for radiometric dating and also clarified certain puzzling stratigraphic-structural problems in the area.

Excavation at the site was carried out by Tedford Construction Company of Fallon, Nevada. A Caterpillar tractor exposed the strata by making two 30-meter cuts along the strike of the beds. Without this initial assistance, a large collection could not have been made.

Abstract

The Buffalo Canyon flora of 69 species, dated at 18.0 Ma, comes from ashy diatomaceous shale in the lower Buffalo Canyon Formation of western Nevada. Geologic mapping shows that the Buffalo Basin was rimmed by low volcanic hills, and that the flora accumulated in its southeast part where the lake was deepest. The fossil plants were transported by turbidity current into the lake in late summer or autumn. Most of the species have allied taxa in the Klamath Mountains of northwestern California and the Cascades to the north. They contribute to mixed conifer forest, with broadleaved evergreen forest on adjacent warmer, drier sites. In addition, a number of taxa have allies in the eastern United States and eastern Asia, regions with ample summer rainfall. The occurrence of numerous fossil conifers allied to modern high montane taxa is attributed to a richer mixed conifer-deciduous hardwood forest resulting from an equable climate and ample summer rainfall. Since the bordering terrain was low, there is no evidence to support the belief that the abundant conifers of upland affinity were transported from distant mountains. Annual precipitation was about 90-100 cm (35-40 in.), distributed chiefly as summer rain and light winter snow. Mean annual temperature was near 10°C, and the mean annual range about 14°C. This represents a warmth of W 12.5°C, with 138 days (4.6 months) warmer than that. Paleotemperature estimates suggest that the lake basin had an elevation near 1,280 m (4,200 ft.). Regional comparisons demonstrate the diversity of Middle Miocene floristic provinces. Early to Middle Miocene (19-15 Ma) vegetation boundaries persisted into the later Middle Miocene (14-13 Ma), when a drastic decrease in summer rainfall resulted in major changes in floral composition in each province. As noted previously (Axelrod, 1985), this probably resulted from the spread of Antarctic ice sheet and its effect on global climate.

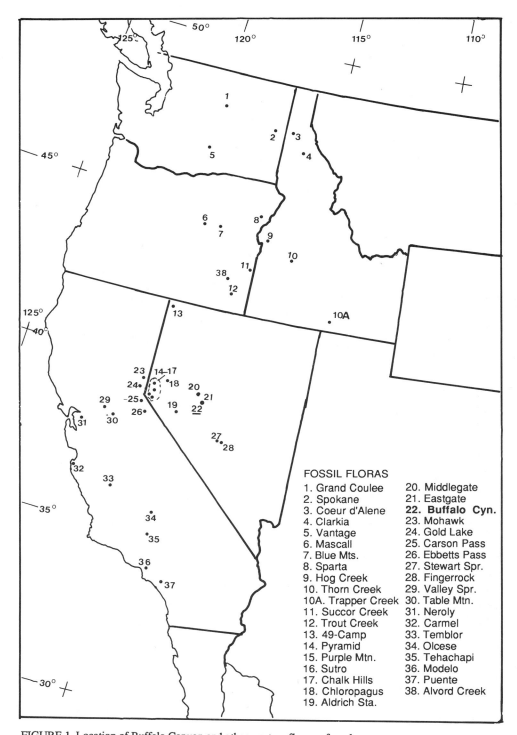

FOSSIL FLORAS

1. Grand Coulee
2. Spokane
3. Coeur d'Alene
4. Clarkia
5. Vantage
6. Mascall
7. Blue Mts.
8. Sparta
9. Hog Creek
10. Thorn Creek
10A. Trapper Creek
11. Succor Creek
12. Trout Creek
13. 49-Camp
14. Pyramid
15. Purple Mtn.
16. Sutro
17. Chalk Hills
18. Chloropagus
19. Aldrich Sta.
20. Middlegate
21. Eastgate
22. **Buffalo Cyn.**
23. Mohawk
24. Gold Lake
25. Carson Pass
26. Ebbetts Pass
27. Stewart Spr.
28. Fingerrock
29. Valley Spr.
30. Table Mtn.
31. Neroly
32. Carmel
33. Temblor
34. Olcese
35. Tehachapi
36. Modelo
37. Puente
38. Alvord Creek

FIGURE 1. Location of Buffalo Canyon and other western floras referred to

INTRODUCTION

The rich, well-preserved Buffalo Canyon flora of 69 species is the oldest (18.0 Ma) conifer-deciduous hardwood forest presently known from the western Great Basin (Fig. 1). To the north, the only flora of comparable age is the Alvord Creek (ca. 21 Ma) of southeastern Oregon (Axelrod, 1944b). It also represents a conifer-hardwood forest that lived in an upland region, probably in a caldera prior to the outpouring of Steens Basalt (ca. 15 Ma), a unit of the Columbia Lava province. To the south, the oldest Neogene flora is the Tehachapi (17.5 Ma) from the southeast end of the Sierra Nevada (Axelrod, 1939). Its taxa are very different, representing members of an evergreen oak-cypress-piñon woodland and arid subtropic scrub vegetation of a separate floristic province at lower elevation in a warmer climate. To the west, the Sutro flora (ca. 21 Ma) near Silver City, Nevada, is dominated by exotic broadleaved evergreens and conifers. They represent a tongue of the rich mesophytic vegetation from the windward slope of the low Sierra Nevada which reached up a broad valley that drained west to the sea, prior to the uplift of the Virginia Range, Carson Range, and Sierra Nevada.

The Buffalo Canyon flora thus provides critical information regarding Early Miocene regional environments at an earlier time than the previously described floras in west-central Nevada. These include the nearby Eastgate and Middlegate floras, dated at 18 Ma (Axelrod, 1985), and the impoverished, younger (14-13 Ma) Aldrich Station, Chloropagus, and Purple Mountain floras (Axelrod, 1956, 1976b). To the south, in the Cedar Mountain area, are the Fingerrock (16.4 Ma) and Stewart Spring (14 Ma) floras described by Wolfe (1964a), with the latter under revision on the basis of a more adequate sample (H.E. Schorn, in prep.). The Buffalo Canyon flora can thus clarify the factors that account for the progressive impoverishment of Neogene floras in the region. In addition, the presently known floras provide data regarding the boundaries between the Great Basin floristic region and those to the north, south, and west. The boundaries (really ecotones) can be estimated in terms of both precipitation and temperature that express the general climate of the floristic provinces. Of more local interest are the causes of the marked differences in composition between the Buffalo Canyon and the essentially contemporaneous Eastgate and Middlegate floras only 19 km (12 mi.) and 27 km (17 mi.) northwest, respectively.

1

PRESENT PHYSICAL SETTING

The Buffalo Canyon flora occurs in the southeast corner of a long, relatively narrow valley bounded by the low Eastgate Hills on the west and the high Desatoya Range at the east. The flora takes its name from its occurrence near the mouth of Buffalo Canyon in the southern Desatoya Range. This range rises to 2,890-3,000 m (9,000-9,500 ft.) in its summit section to the north, whereas the Eastgate Hills to the west have an average summit level near 1,975-2,130 m (6,500-7,000 ft.). The valley between them increases in elevation from near 1,670 m (5,500 ft.) in the north to 1,975 m (6,500 ft.) in the south, where it terminates against volcanic hills. Drainage is west from the Desatoya Range, through a deep gorge that cuts across the Eastgate Hills to emerge at Eastgate, thence across the desert in the Middlegate Basin, through the south end of the Clan Alpine Range, and then north to the sink of Dixie Valley, some 120 km (75 mi.) distant.

The fossil locality is in the lower piñon-juniper woodland belt (Plate 1). The low trees, *Pinus monophylla* and *Juniperus osteosperma*, are associated with diverse shrubs, such as *Artemisia tridentata, Cercocarpus ledifolius, Chrysothamnus nauseosus, Ephedra viridus*, and *Purshia tridentata*. Streamways in the Desatoya Range are lined with *Alnus tenuifolia, Populus tremuloides* and shrubby species of *Amelanchier, Prunus, Ribes, Rosa* and *Salix* which fill the narrow canyon bottoms. To the west, the stream through the Middlegate basin has greasewood (*Sarcobatus vermiculatus*) and an occasional clump of Fremont cottonwood (*Populus fremontii*). The bordering plain has a typical Great Basin desert flora of *Artemisia confertifolia, A. spinosus, Ephedra nevadensis*, and *Lycium cooperi*.

The area has a continental climate, with hot summers and cold winters. The nearest meteorological station was at Eastgate (now abandoned), where an 8-year record gives an indication of conditions. Mean annual temperature there is about 10.8°C (51.5°F) and the annual range is 23.3°C (42°F). The fossil locality is 15 km southeast (9 mi.) and ca. 300 m (1,000 ft.) higher, so it has cooler summers and winters. The growing season (W=warmth) at Eastgate has 158 days warmer than W 13.2°C (55.7°F). Since the fossil locality is ca. 300 m (1.6°C) higher, the growing season there is ca. 30 days shorter, i.e., W 12.2°C, or 128 days warmer than 12.2°C as measured on the nomogram (see Fig. 3). At Eastgate, extreme low winter temperatures reach -8°C (17°F), and the summers have highs of 39.4°C (103°F). They reach into the middle

30s°C in the piñon belt where the flora occurs. Precipitation at Eastgate, at the edge of the desert, is 125 mm (5 in.) but rises to near 380-450 mm (15-18 in.) in the lower piñon-juniper belt at the fossil site. Precipitation comes chiefly as rain and snow from late fall to spring. Occasional summer showers from tropical air masses that range north may result in local heavy rain ("cloudbursts"), but regionally they add little moisture to the soil, except over the higher basin-ranges to the east.

GEOLOGY

A geologic reconnaissance of the Desatoya region is included in a report on the geology of Churchill County (Willden and Speed, 1974). A detailed report on the geology of the southern Desatoya Mountains and the bordering Eastgate Hills was prepared by Barrows (1971, see Fig. 2, in pocket). The volcanic rocks that make up the bulk of the Desatoya Mountains and adjacent Eastgate Hills are approximately 3,000 m thick. They are composed largely of rhyolite and quartz latite flows, welded tuffs, and ash-flow tuffs that range in age from 31 to 23 Ma as judged from radiometric dating (Barrows, 1971). They rest on a basement of Trias-Jurassic marble, schist, and slate that has been highly folded and faulted and intruded locally by granodiorite of presumed Cretaceous age. The nature of the volcanic formations on which the Buffalo Canyon Formation rests is outlined in Table 1.

BUFFALO CANYON FORMATION

This formation, which rests unconformably on the older volcanic formations, is made up of lacustrine, volcanic, and deltaic sedimentary rocks. It is exposed in the lowland area between the southern Desatoya Mountains and Eastgate Hills, chiefly south of State Highway 2 (old U.S. 50). The formation thins in the northern part of the basin, where it laps onto quartz latite welded tuffs of the Desatoya Formation, or onto the massive red-brown latite ashflows of the Skull Formation. In the southern part of the basin it rests on andesite flows of the Tortoise Peak Formation and to the southwest it interfingers with the basal tuffs of the Buffalo Hill Volcanics, chiefly latite pyroclastics. Since the sedimentary rocks of the Buffalo Canyon Formation are weakly indurated, much of it has been carved into badlands. The formation is about 475 m (1,560 ft.) thick and includes several members (Barrows, 1971).

1. Ashflow member

This unit forms the basal part of the formation and is best exposed in the southeastern part of the basin where it rests unconformably on all the older volcanic rocks (Table 1). The member is fully exposed at the mouth of Buffalo Canyon, where vitric ashflow and ashfall tuffs are interbedded with pyroclastic water-worn ejecta, mudstone, chert, and diatomaceous silts and claystone. The member thickens to the

TABLE 1

Stratigraphic Units in the Buffalo Canyon Area

AGE Ma.*	FORMATION	THICK-NESS (m)	GENERAL LITHOLOGY
<1	PLEISTOCENE	45	Recent alluvium, terraces, older alluvium, all poorly indurated.
18.0	Buffalo Canyon	475	Mem. 5. Fanglomerate-mudflow with salmon mudstone exposed only to the northeast. Mem. 4. Salmon mudstone, ss, pebble conglomerate. Mem. 3. Light-brown mudstone, ss, pebble beds, vitric tuffs. Mem. 2. Diatomite, some silt stone, ss, and thin vitric tuffs. Mem. 1. Latite ashflow tuff, pumiceous, grading up to mud-stone, chert, diatomite. Ash-flow tuff (18.0 Ma) thickens to the north.
20.0	Erb	75	White to pink latite welded vitric ashflow tuffs, with pebble cgl. at base.
21.9	Tortoise Peak	90	Mem. 1. Dark gray to black porphyritic hypersthene ande-site flows, breccias. Mem. 2. Epiclastic and pyroclas-tic yellow-brown andesites, tuffs, ss, shales.
23	Desatoya	900	Silicic ashflow tuffs, numerous cooling units, densely welded silicic quartz latite, vitric ashflow tuffs. Dark gray-brown, forms steep cliffs.
25	Skull	120	Devitrified red-brown to black trachytic vitric ashflow tuff, platy and blocky, very dense and hard, forms steep cliffs.
27?	Carroll	600	White, pink, gray vitric crystal rhyolite tuff and ashflow tuffs. Local ashflow tuffs and fine-grained sediments. Local qtz. diorite and diorite clasts, some very large.
>70	Plutonic Rocks	?	Slide blocks of qtz. diorite, qtz. monzonite, altered greenish diorite in volcanics.

*K/Ar ages in Barrows (1971), recalculated using constants of Dalrymple (1979). Buffalo Canyon Formation re-dated (3 dates; see Appendix).

north and locally rests on a reddish-brown sediment that appears to be a paleosoil on the Erb Formation. To the south and west it is replaced by the Diatomite member which overlaps the volcanic rocks. The ashflow varies 3 to 60+ m thick.

The ashflow tuffs range from about 5 m to 30 m thick and are white or, yellow-brown to gray. The volcanic debris includes glass shards and ash-size fragments, as well as crystals and pumice fragments. Compressed vitriclastic fragments are common in the tuff. Resting on the vitric tuff are mudstone, thin tuff, and diatomite, as well as chert that appears to be a silicified tuff, judging from the progressive change upward from fresh vitric tuff to porcellanite to chert. The alkali feldspars in 3 ashflow tuffs in the lower Buffalo Canyon Formation yielded an average K/Ar of 18 Ma (see Appendix). As the vitric tuffs decrease at higher levels, diatomite increases and passes upward conformally into the Diatomite member.

2.   Diatomite member

This unit, about 90 m thick, is composed of dominantly white diatomite and interbedded buff-colored diatomaceous mudstone, siltstone, shale, and numerous fine pumiceous beds. Locally in the basin, as at the mouth of Buffalo Canyon, the basal bed is a conglomerate about 3 m thick composed of rounded to subrounded cobbles of locally derived volcanic rocks in a sandy matrix. It rests on the ashflow tuff member and thins basinward. The conglomerate evidently was deposited opposite a valley in volcanic terrain to the east. To the south and east, the member rests on weathered dark andesites of the Tortoise Peak Formation. To the north and west it grades up into the Mudstone member.

The Diatomite member is composed largely of nearly pure, white to light-brown diatomite and diatomaceous earth interbedded with moderately indurated diatomaceous and tuffaceous siltstones and gray to white pumiceous silvery tuffs. The diatomite may occur in very thin laminated beds interlayered with siltstone, sandstone, or tuff, or in massive beds that show no obvious sedimentary structures. The siltstone and shale beds are from 1.5 to 10 cm thick, and the sandstone beds are usually 20 to 30 cm thick. The beds are moderately indurated by compaction. The siltstones and sandstones have a high volcanic glass content, and diatoms are present. The volcanic glass occurs in the form of pumice fragments, shards, and dust. Some tuff beds are up to 4 m thick and show a variety of sedimentary features, notably graded bedding, cut-and-fill, and slump structures.

The impressions of leaves and winged seeds that make up the Buffalo Canyon flora occur in diatomaceous shales in the upper part of the member. The site is on a south-facing slope where soil is thin and vegetation sparse. At the locality the beds dip 15° to 18° west.

3.   Brown mudstone member

This unit conformably overlies the Diatomite member and forms a sequence of buff-colored, poorly lithified, fine-grained epiclastic and pyroclastic rocks in the south-central part of Buffalo Basin. To the east and northeast it rests unconformably on, and is in fault contact with, welded latite tuffs of the Skull Formation. On the west and southwest sides of the basin it rests unconformably on, and is also in fault contact with, quartz latite welded tuffs of the Desatoya Formation and andesites of the Tortoise Peak Formation. To the north it interfingers laterally with the Salmon mudstone member.

Much of the area covered by the Brown mudstone member has been eroded into a spectacular badland terrain in the major drainages north of Buffalo Canyon. This member is about 275 m thick, but thins as it laps onto the volcanic rocks that surround and floor the basin.

Brown to buff mudstone beds make up about 65 percent of this member. They vary from 15 cm to 1.5 m thick and are interbedded with tuffaceous sandstone, siltstone, and shale of varied thickness and extent. Near the top of the member are thin beds of glassy tuff, pebbly sandstone, and pebble conglomerate. The pebbles are subrounded to rounded plutonic and volcanic rock fragments from the east side of the basin. The coarsest fraction occurs chiefly in the southeastern part of Buffalo Basin in the upper part of the member, generally in beds up to 1 to 2 m thick. In some of the beds layers of volcanic glass shards and pumice fragments are abundant. Ashfall layers may be 20 to 25 cm thick and are white or light gray to silvery and poorly indurated.

4.   Salmon mudstone member

Largely concealed by Quaternary alluvial deposits, this member occurs in the central, western, and northern parts of the basin. It is best exposed in badlands along the county road that leads south through the area. This member is made up chiefly of lacustrine sediments that rest conformably on the underlying Brown mudstone member. The Salmon mudstone member becomes conglomeratic in the northeast part of Buffalo Basin. In the northern part of the basin outside of the map in Fig. 2 this member laps onto topographic highs of welded tuffs of the Desatoya Formation and onto the Skull Formation on the east side of the basin adjacent to Highway 2. The higher parts are concealed by alluvial fans from the Eastgate Hills and Desatoya Range, so its total thickness cannot be determined, but it is certainly in excess of 200 m.

This member consists mainly of alternating beds of salmon-colored, gray-white, and light brown mudstone and tuff that make up about 95 percent of the unit. The remainder consists of lenses of siltstone, sandstone, and pebble to small-cobble conglomerate. Many of the mudstone beds have a high volcanic glass component and also contain clay minerals of the montmorillonite group as well as diatom frustules. The pebbly sandstone and conglomerate contain clasts similar to rocks that crop out in the Eastgate Hills to the west, showing that they formed the west wall of the basin during deposition of the Buffalo Canyon Formation.

5.   Fanglomerate-mudflow member

A thick, wedge-shaped deposit of poorly indurated conglomeratic salmon mudstone with sandstone and gravel lenses occurs in gullies bordering State Highway 2 in the northeastern part of the basin, and extends to the south. The conglomeratic mudstone is massive, and the tuffaceous sandstones exhibit graded bedding, cross-cutting, and other fluvatile features. The conglomeratic portions consist of tan to salmon-colored mudstone with angular and subangular pebbles, cobbles, boulders, and blocks derived chiefly from the underlying Skull Formation at the east. Clasts from the Carroll Formation, chiefly welded rhyolite tuffs that crop out in the middle and upper parts of the Desatoya Range to the east, occur higher in the section. These coarser fractions were carried by a relatively large river from hills at the present site of the Desatoya Mountains at the northeastern margin of Buffalo Basin before the principal uplift of the present range. The coarse conglomeratic facies is overlain by an old alluvial fan deposit that dips gently west.

PHYSICAL SETTING

The epiclastic material in the Buffalo Canyon Formation came from the hilly terrain surrounding the present basin. Clasts in the conglomerates and coarse sandstones of each member represent rock types now exposed around the margins of the basin. The clasts in many of the coarse-grained deposits are angular to subrounded, indicating transport for a relatively short distance. It is only in the conglomeratic facies of the Salmon mudstone member that a somewhat greater distance of transport is implied by the rounding of the clasts. Their source was higher up in the ancestral Desatoya Range. The Salmon mudstone member also contains coarse volcanic clasts from the Eastgate Hills indicating that they were elevated. This agrees with evidence in the adjacent Middlegate Basin to the west, where thick sedimentary breccias from the western front of the Eastgate Hills are interbedded in the Middlegate Formation of similar age. At present, no evidence has been found that suggests a major external source for the coarse clasts in the formation.

The evidence implies that Buffalo Basin was formed by subsidence following eruption of a major volcanic field. The general area of the Desatoya Mountains, the Clan Alpine Mountains to the west, and the Shoshone Mountains to the east has been depicted as composed of chaotic units in a major volcano-tectonic depression (Burke and McKee, 1979). That Buffalo Basin was part of this depression seems reasonably certain. The formations in the volcanic sequence below the Buffalo Canyon Formation are of only local occurrence, implying eruption within a tectonic depression. This is suggested also by the local occurrence in the volcanic terrain on the east side of the basin of slide blocks of granodiorite and metamorphic rocks. Exotic blocks have been mapped also in the southeast corner of the Middlegate Basin (Axelrod, 1985, see map), and are nicely illustrated for the Northumberland caldera (McKee, 1974, fig. 8).

The composition of the several members of the Buffalo Canyon Formation shows that there was active volcanism during deposition. In the southwestern and southern part of the basin field evidence shows that ashfalls, derived chiefly from eruptions of glassy tuffs, interfinger with the Diatomite and Brown mudstone members (Barrows, 1971).

Buffalo Basin, in which the Buffalo Canyon Formation occurs, came into existence after collapse of the volcanic formations that now form hills bordering the basin and are fully 3,000 m thick. Displacements along faults, and extrusion of rhyolite and latite ash-flows and ashfalls of the Ash-flow tuff member from small vents in the southern part of the area, appear to have blocked drainage to the south, thus largely closing the basin about 20 Ma ago. The nature of terrain surrounding the lake is suggested by the sedimentary record. Relief was low to the east, south and west of the lake, slopes were gentle, and most streams entering the lake probably were small and of low gradient. This is inferred because the epiclastic sediments in those parts of the basin are chiefly mudstone, and most of the clastic sediments are not coarse. Several kilometers north of Buffalo Canyon, on the northeast side of the lake, slopes were steeper and relief moderate, as shown by the coarse Fanglomerate member. Its large clasts of welded tuffs were transported by a major stream that drained the central, ancestral Desatoya Range.

Judging from the distribution of the members, Buffalo Lake initially occupied the deepest part of the basin in its southeastern part where the Ashflow and Diatomite members accumulated. The lake gradually expanded northward, filling the basin with sediments of the younger members. Locally, there are rich reed-bearing layers in the mudstones, suggesting that at times the lake had shallow and swampy shores. The presence of diatom remains throughout the formation implies that the water was cool, clear, and well lighted, for the most part; that it was rich in phosphate, nitrate, and silica; and that it had a generally neutral pH. At the fossil site, where many cubic meters of shale were split to recover fossil plants, occasional remains of fish were also recovered. This implies that the lake was not in a closed basin during its early history. However, to judge from the upper members, which have numerous beds of coarse clasts derived from the bordering hills, the basin may have been closed later in its history as relief increased.

TAPHONOMY

The fossil leaves are well preserved and show little or no evidence of destruction during transport to the site. Further, it does not appear that they lay on the ground or in pools for any length of time, for they are not decomposed to any degree; this implies rapid burial. Most leaves must have accumulated in autumn, consistent with the abundance of winged seeds of spruce, pine, and hemlock, for their cones ripen and seeds are shed in late summer to early autumn. The fact that many of the leaves have serrate margins implies a mild to cool temperate climate, which is also consistent with leaf fall in autumn. As will be discussed, nearly all of the taxa are allied to species living in temperate regions.

The fossil site is at least two km distant from the shore. The leaves and winged seeds are concentrated within a foot or so in the diatomite, and their abundance decreases rapidly along strike and vertically. The diatomite section shows finely graded bedding at the locality, as well as slump structures and concurrent erosion. In addition, occasional small volcanic pebbles, most about 0.5 cm long, occur in the diatomite. These data suggest transport by turbidity current at a time of high input of water into the lake, possibly during a severe autumn storm. Such a storm, accompanied by high winds, would provide a means for stripping leaves ready to fall and shaking winged seeds from cones, which would then rain down into the lake-border area near an entering stream. One need only examine concentrations of leaves in a park or forest following a period of high winds in autumn to appreciate that leaves and winged seeds are regularly transported moderate distances — some for tens of meters — and swept into local accumulations. Following such transport to the lake-margin area, they would settle into the water and could thence be transported by turbidity current out into the lake. As the current waned, the structures in suspension would be deposited on the lake floor and rapidly buried.

Distance from the shore clarifies the representation of taxa. The flora is dominated by the hard, durable leaves of *Quercus hannibali*, which would readily withstand transport if suspended in a current that entered the lake. Its abundance also indicates that it dominated an evergreen sclerophyll forest (or woodland), probably on warmer south-facing slopes, much as its modern descendant *Q. chrysolepis* does today. Birch leaves and would easily be carried in suspension because of their large surface area and light weight. The stream- and lake-border habitat of birches no doubt facilitates their abundant contribution to the record. This also accounts for the abundance of other riparian taxa, notably the 7 species of *Salix* and 3 of *Populus*. They probably formed dense deciduous communities along the lake shore and entering streams and were thus in a favorable position for currents to transport their leaves into the lake.   The abundant leaves of *Zelkova* imply that it was also a common member of the riparian and lake-border deciduous vegetation.   The winged seeds of conifers, especially those of *Larix, Picea, Pinus*, and *Pseudotsuga*, are also abundant.   They would readily float on the surface and would also be carried by current into the lake for some distance before they became waterlogged.   The absence of remains of cones, and the rarity of conifer needles and twigs, is consistent with their high area-weight ratio: they were deposited nearer the shore, for the most part.   A notable exception is the relative abundance of *Juniperus* twigs.   Juniper trees probably lived on the drier, west-facing slopes of hard, well-drained volcanic rocks bordering the lake, where they were associated with live oak, madrone, and other taxa of subhumid requirements.   These and juniper were thus in a favorable position to contribute fairly abundantly to a record deposited by turbidity currents.

Another obvious factor that accounts for the relative representation of taxa in the flora is size of the plant.   Clearly, a large tree growing along the edge of a stream entering the lake would contribute more material to an accumulating record than a small shrub on the forest floor a few yards from a stream or the lake shore.   In this regard, tree species outnumber shrubs 38 to 29.   More important, however, is the numerical abundance of their specimens, with 6,119 representing trees, whereas only 1,232 come from shrubs and small trees, a ratio of 80% to 20%.   Since *Typha* and *Nymphaeites* were not included in these counts, the above total (7,351 specimens) is less than that (8,276 specimens) for the entire flora.

To summarize, the fossil flora occurs in a small, generally oblong basin surrounded by Oligocene-Lower Miocene volcanic rocks 3,000 m thick.   The local distribution of the welded tuffs and flows suggests that the basin is part of a major tectonic-collapse structure, consistent with the occurrence of slide blocks of granodiorite and metamorphic rocks in the volcanic sequence bordering Buffalo Basin.

The fossil flora occurs in diatomite conformably 80 m above tuffs dated at 18 Ma.   Assuming a normal rate of deposition, the flora probably is not much younger.   The Diatomite member accumulated in the deepest part of the basin, in its southeastern corner.   Fish remains in this member indicate that the basin was not closed initially, although the increasingly coarse clasts in the upper members derived from bordering hills suggest that it may have been closed later in its history.   Surrounding hills were relatively low, judging from the small to medium clasts in the rare conglomerates, though in the northeastern sector the upper member has large clasts derived from higher terrain there.

The fossil leaves and winged seeds occur in diatomite associated with finely graded ashy beds well offshore.   Current direction and cut-and-fill structures suggest transport by turbidity current into the lake.   The current may have been triggered by a major rainstorm and accompanying high winds in late summer or autumn that loosened leaves and winged seeds from trees and shrubs.   Those that accumulated in an entering stream were thence carried in a dense, ash-laden current well offshore to settle at the fossil locality.

COMPOSITION OF THE FLORA

This flora of 8,276 specimens is represented by especially well-preserved leaves, winged conifer seeds, and samaras of maple, all forming attractive red-brown imprints on the buff diatomaceous shale. The sample is considered representative, since only two additional species, single leaves of *Cercocarpus* and *Chamaebatia*, were recovered during the last two days of a total of 28 man-days digging. The 69 species are distributed among 22 families and 42 genera; 15 species are described as new, and 1 is given a new name. There are 12 conifers, 1 monocotyledon, and 55 dicotyledons. In terms of the number of species, the largest families are the Pinaceae with 10; Rosaceae, 11; Salicaceae, 10; Betulaceae, 5; and Aceraceae, 5 species. The genera with the largest number of species are *Salix* with 7, *Acer* with 5, *Betula, Populus*, and *Ribes* with 4 each, and *Picea* and *Prunus* with 3 each.

SYSTEMATIC LIST OF SPECIES

Pinaceae
>   *Abies concoloroides* Brown
>   *Abies laticarpus* MacGinitie
>   *Larix churchillensis* Axelrod, n. sp.
>   *Picea lahontense* MacGinitie
>   *Picea magna* MacGinitie
>   *Picea sonomensis* Axelrod
>   *Pinus balfouroides* Axelrod
>   *Pinus ponderosoides* Axelrod
>   *Pseudotsuga sonomensis* Dorf
>   *Tsuga mertensioides* Axelrod

Cupressaceae
>   *Chamaecyparis cordillerae* Edwards and Schorn
>   *Juniperus desatoyana* Axelrod, n. sp.

Typhaceae
>   *Typha lesquereuxii* Cockerell

11

Salicaceae
  Populus cedrusensis Wolfe
  Populus eotremuloides Knowlton
  Populus payettensis (Knowlton) Axelrod
  Populus pliotremuloides Axelrod
  Salix churchillensis Axelrod, n. sp.
  Salix desatoyana Axelrod
  Salix laevigatoides Axelrod
  Salix owyheeana Chaney and Axelrod
  Salix pelviga Wolfe
  Salix storeyana Axelrod
Betulaceae
  Alnus latahensis Axelrod, n. sp.
  Betula ashleyii Axelrod
  Betula desatoyana Axelrod, n. sp.
  Betula idahoensis Smith
  Betula thor Knowlton
Carpinaceae
  Carpinus oregonensis Axelrod, new name
Juglandaceae
  Carya bendirei (Lesq.) Chaney and Axelrod
  Juglans desatoyana Axelrod, n. sp.
Fagaceae
  Chrysolepis sonomensis Axelrod
  Quercus hannibalii Dorf
  Quercus wislizenoides Axelrod
Ulmaceae
  Ulmus speciosa Newberry
  Zelkova brownii Tanai and Wolfe
Berberidaceae
  Mahonia macginitiei Axelrod
  Mahonia reticulata (MacGinitie) Brown
Nymphaeaceae
  Nymphaeites nevadensis (Knowlton) Brown
Hydrangeaceae
  Hydrangea bendirei (Ward) Knowlton
Grossulariaceae
  Ribes barrowsae Axelrod, n. sp.
  Ribes bonhamii Axelrod, n. sp.
  Ribes stanfordianum Dorf
  Ribes webbii Wolfe

Rosaceae
 Amelanchier desatoyana Axelrod, n. sp.
 Cercocarpus ovatifolius Axelrod
 Chamaebatia nevadensis Axelrod, n. sp.
 Crataegus middlegatei Axelrod
 Heteromeles desatoyana Axelrod, n. sp.
 Lyonothamnus parvifolius (Axelrod) Wolfe
 Prunus chaneyii Condit
 Prunus moragensis Axelrod
 Prunus treasheri Chaney
 Rosa harneyana Chaney and Axelrod
 Sorbus cassiana Axelrod
Fabaceae
 Amorpha stenophylla Axelrod, n. sp.
 Robinia bisonensis Axelrod, n. sp.
Aceraceae
 Acer medianum Knowlton
 Acer negundoides MacGinitie
 Acer oregonianum Knowlton
 Acer trainii Wolfe and Tanai
 Acer tyrrellii Smiley
Meliaceae
 Cedrela trainii Arnold
Ericaceae
 Arbutus trainii MacGinitie
Vacciniaceae
 Vaccinium sophoroides (Knowlton) Brown
Myrtaceae
 Eugenia nevadensis Axelrod
Oleaceae
 Fraxinus desatoyana Axelrod, n. sp.
 Fraxinus eastgatensis Axelrod, n. sp.
Caprifoliaceae
 Symphoricarpos wassukana Axelrod

The quantitative representation of specimens shows that taxa that regularly occur on stream and lake borders contributed abundantly to the flora (Table 2). Among these are species of *Acer, Betula, Juglans, Populus, Salix,* and *Zelkova,* whose remains account for a total of 4,139 specimens, or 50% of the flora. The abundant leaves of *Quercus hannibalii,* which make up another 26% of the specimens, may be attributed to its occurrence on warm south- and west-facing slopes bordering the lake. In this regard, *Q. chrysolepis,* which has leaves and acorn cups similar to those of *Q. hannibalii,* regularly forms dense, pure stands well up into the mixed conifer forest on warm south- and west-facing slopes, and it also frequents stream banks.

Shrubs with a relatively high representation of leaves include *Salix churchillensis* (53 specimens), *S. pelviga* (52), *S. storeyana* (424), *Betula ashleyii* (26), *Sorbus* (31), *Mahonia reticulata* (23), *M. macginitiei* (14), *Prunus* (12), and *Amelanchier* (12), altogether totaling 7.8% of the specimens collected. These are chiefly forest taxa that prefer moist sites along stream and lake margins, and hence were in a favorable position for entry into the record. Their structures are not so abundant as those of the trees, however, since they are lower in stature and produce fewer leaves to contribute to an accumulating record.

The relative abundance of small winged conifer seeds of *Larix, Picea, Pinus, Pseudotsuga,* and *Tsuga,* with a total of 685 specimens (8%), is largely the result of transport by water and wind during autumn. Although the conifers may have lived near the shore, especially along entering streams, the contribution of conifer needles or cones to an accumulating deposit well offshore would be quite limited, since their high area-weight ratio as compared with winged seeds would result in their deposition near the shore. The abundant representation of leaves of alder, birch, willow, cottonwood, and walnut, which probably formed a dense deciduous woodland bordering the lake, as well as the abundance of evergreen oak and madrone on nearby warmer slopes, suggests that the mixed conifer-hardwood forest probably occupied cooler north- and east-facing slopes away from the immediate shore area, though some trees were scattered there.

TABLE 2
Number of Identified Specimens in the Buffalo Canyon Flora*

Quercus hannibalii		2,191
leaves	2,187	
acorns	4	
Betula thor		2,128
leaves	2,123	
seeds	4	
catkin	1	
Typha lesquereuxi		795
Zelkova brownii		744
Salix storeyana		424
Betula desatoyana		390
Picea lahontense		272
winged seeds		
Picea sonomensis		227
winged seeds		
Acer negundoides		122
samaras	114	
leaflets	8	
Juniperus desatoyana	117	
leafy twigs	116	
berry	1	
Populus cedrusensis		74
Arbutus prexalapensis		68
Juglans desatoyana		56
Salix churchillensis		53
Salix pelviga		52
Tsuga mertensioides		48
winged seeds		
Ulmus speciosa		42
leaves	6	
samaras	36	
Pseudotsuga sonomensis		40
winged seeds	38	
needles	2	
Picea magna (winged seeds)		40
Sorbus cassiana		31
Fraxinus desatoyana		29
samaras	26	
leaflets	3	

Betula ashleyi		26
Pinus ponderosoides		23
Mahonia reticulata		23
Pinus balfouroides		21
winged seeds	16	
fascicles	4	
branchlet	1	
Populus pliotremuloides		18
Populus payettensis		17
Larix churchillensis		14
winged seeds		
Mahonia macginitiei		14
Prunus treasheri		12
Amelanchier desatoyana		12
Carya bendirei		11
Populus eotremuloides		10
Lyonothamnus parvifolius		10
Quercus wislizenoides		9
Rosa harneyana		9
Nymphaeites nevadensis		8
leaves	3	
root scars	5	
Chrysolepis sonomensis		8
Acer oregonianum		7
samaras		
Heteromeles desatoyana		7
Prunus chaneyii		6
Salix owyheeana		6
Salix laevigatoides		5
Fraxinus eastgatensis		5
leaflet	1	
samaras	4	
Cedrela trainii		4
Salix desatoyana		3
Ribes webbi		3
Crataegus middlegatei		3
Eugenia nevadensis		3
Hydrangea bendirei		3
Ribes barrowsii		3
Abies concoloroides		2
winged seeds		
Abies laticarpus		2
winged seeds		
Ribes bonhamii		2
Amorpha stenophylla		2
Symphoricarpos wassukana		2
Alnus latahensis		2
Chamaecyparis cordillerae		1
Acer medianum		1
Betula idahoensis		1
Ribes stanfordianum		1
Cercocarpus ovatifolius		1
Chamaebatia nevadensis		1
Prunus moragensis		1
Robinia bisonensis		1
Acer trainii (samara)		1
Acer tyrrellii (samara)		1
Vaccinium sophoroides		1

Total: 8,276

*These fossils occur along strike for about 40 m and are distributed stratigraphically for about 1 m.

ALLIED MODERN VEGETATION

Table 3 lists modern species allied to the fossil taxa and indicates their present areas of occurrence. Most of the fossil species have similar living taxa in the western United States, with fewer in the eastern United States or eastern Asia, as discussed below. Judging from the habit of these modern species, the flora includes 37 trees, 30 shrubs or small trees, and 2 aquatic herbaceous perennials (Table 4). The trees include 12 conifers, of which 11 are evergreen and 1 (*Larix*) is deciduous. Of the 25 dicotyledons, 19 are deciduous and 6 evergreen. Among the 30 shrubs or small trees, there are 25 deciduous and 5 evergreen species, as judged from the habit of allied modern taxa.

TABLE 3

Distribution of Modern Species Allied to Buffalo Canyon Taxa

Fossil Species	Western U.S.	Eastern U.S.	Eastern Asia
Abies concoloroides	*A. concolor*		
Abies laticarpus	*A. shastensis*		
Larix churchillensis	*L. occidentalis*		
Picea lahontense			several spp.
Picea magna			*P. polita?*
Picea sonomensis	*P. breweriana*		
Pinus balfouroides	*P. balfouriana*		
Pinus ponderosoides	*P. ponderosa*		
Pseudotsuga sonomensis	*P. menziesii*		
Tsuga mertensioides	*T. mertensiana*	*T. caroliniana*	
Chamaecyparis cordillerae	*C. lawsoniana*		
Juniperus desatoyana	*J. occidentalis*	*J. virginiana*	
Typha lesquereuxii	*T. latifolia*	*T. latifolia*	*T. latifolia*
Populus cedrusensis (N. Mexico)	*P. brandegeei*		
Populus eotremuloides	*P. trichocarpa*	*P. balsamifera*	*P. adenopoda*
Populus payettensis	*P. angustifolia*		
Populus pliotremuloides	*P. tremuloides*	*P. tremuloides*	*P. tremula*
Salix churchillensis	*S. exigua*		
Salix desatoyana	*S. nigra*		
Salix laevigatoides	*S. laevigata*		
Salix owyheeana	*S. hookeriana*		
Salix pelviga	*S. melanopsis*		
Salix storeyana	*S. lemmonii*		
Alnus latahensis		*A. maritima*	*A. japonica*
Betula ashleyi	*B. fontinalis*		
Betula desatoyana		*B. cordifolia*	
Betula idahoensis		*B. lenta*	
Betula thor		*B. papyrifera*	
Carpinus oregonensis		*C. caroliniana*	*C.* spp.
Carya bendirei		*C. ovata*	
Juglans desatoyana	*J. major*		
Chrysolepis sonomensis	*C. chrysophylla*		
Quercus hannibalii	*Q. chrysolepis*		
Quercus wislizenoides	*Q. wislizenii*		
Ulmus speciosa		*rubra?*	
Zelkova brownii			*Z. serrata*
Mahonia macginitiei	*M. aquifolium*		
Mahonia reticulata	*M. insularis*		
Nymphaeites nevadensis	*Nuphar polysepalum*	*N.* spp.	*N.* spp.
Hydrangea bendirei			*H. aspera*
Ribes barrowsae	*R. aureum*		
Ribes bonhamii		*R. americanum*	*R. meyeri*

Ribes stanfordianum	*R. nevadensis*		
Ribes webbii	*R. cereum*		
Amelanchier desatoyana	*A. florida*		
Cercocarpus ovatifolius	*C. blancheae*		
Chamaebatia nevadensis	*C. foliolosa*		
Crataegus middlegatei	*C. chrysocarpa*		
Heteromeles desatoyana	*H. arbutifolia*		
Lyonothamnus parvifolius	*L. asplenifolius*		
Prunus chaneyi	*P. demissa*	*P. virginiana*	*P. wilsonii*
Prunus moragensis	*P. emarginata*		
Prunus treasheri			*P. davidiana*
Rosa harneyana	*R. nutkana*	*R.* spp.	
Sorbus cassiana			*S. pohuashanensis*
Amorpha stenophylla		*A. angustifolia*	
Robinia bisonensis		*R. pseudoacacia*	
Acer medianum	extinct		
Acer negundoides	*A. negundo*	*A. negundo*	*A. henryii*
Acer oregonianum	*A. macrophyllum*		
Acer trainii	*A. glabrum*		
Acer tyrrellii	*A. grandidentatum*	*A. saccharum*	
Cedrela trainii (Mexico)	*C. mexicana*		
Arbutus trainii	*A. xalapensis*		
Eugenia nevadensis	Mexican spp.		
Fraxinus desatoyana	*F. velutina*		
Fraxinus eastgatensis			*F. chinensis*
Vaccinium sophoroides		*V. arboreum*	
Symphoricarpos wassukana	*S. oreophilus*		

TABLE 4

Probable Growth Habit of Buffalo Canyon Species

TREES (37)

Conifers (12)

Abies concoloroides	*Pinus balfouroides*
Abies laticarpus	*Pinus ponderosoides*
Larix churchillensis	*Pseudotsuga sonomensis*
Picea lahontense	*Tsuga mertensioides*
Picea magna	*Chamaecyparis cordillerae*
Picea sonomensis	*Juniperus desatoyana*

Dicotyledons (25)

Deciduous (19)

Populus eotremuloides	*Ulmus speciosa*
Populus cedrusensis	*Zelkova brownii*
Populus payettensis	*Robinia bisonensis*
Populus pliotremuloides	*Acer medianum*
Betula desatoyana	*Acer negundoides*
Betula idahoensis	*Acer oregonianum*
Betula thor	*Acer tyrrellii*
Carpinus oregonensis	*Fraxinus desatoyana*
Carya bendirei	*Fraxinus eastgatensis*
Juglans desatoyana	

Evergreen (6)

Chrysolepis sonomensis
Quercus hannibalii
Quercus wislizenoides

Eugenia nevadensis
Cedrela trainii
Arbutus trainii

SHRUBS AND SMALL TREES (30)

Deciduous (25)

Salix churchillensis
Salix desatoyana
Salix laevigatoides
Salix owyheeana
Salix pelviga
Salix storeyana
Alnus latahensis
Betula ashleyii
Hydrangea bendirei
Ribes barrowsae
Ribes bonhamii
Ribes stanfordianum
Ribes webbii

Amelanchier desatoyana
Cercocarpus ovatifolius
Crataegus middlegatei
Prunus chaneyii
Prunus moragensis
Prunus treasheri
Rosa harneyana
Sorbus cassiana
Acer trainii
Amorpha stenophylla
Symphoricarpos wassukana
Vaccinium sophoroides

Evergreen (5)

Mahonia macginitiei
Mahonia reticulata

Lyonothamnus parvifolius
Heteromeles desatoyana
Chamaebatia nevadensis

AQUATIC HERBACEOUS PERENNIALS (2)

Nymphaeites nevadensis

Typha lesquereuxii

Nearly all of the fossil species are similar to living plants. Hence, an examination of present-day vegetation in which these allied modern taxa occur provides a basis for inferring the nature of the vegetation, climate, and elevation of Buffalo Basin in the Early Miocene.

Many of the fossil species are similar to taxa in forests of the Klamath Mountain region of northwestern California. This area is especially notable for the numerous relict and disjunct taxa that occur there (Whittaker, 1960; Stebbins and Major, 1965; Sawyer and Thornburgh, 1977). Represented in this area are the following species with closely related taxa in the fossil flora, as well as several others (in parentheses) that are more distantly related:

Abies concolor
Abies magnifica
Picea breweriana
Picea (engelmannii)
Pinus balfouriana
Pinus ponderosa
Pseudotsuga menziesii
Tsuga mertensiana
Chamaecyparis lawsoniana
Juniperus occidentalis
Populus tremuloides
Populus trichocarpa

Mahonia aquifolium
Mahonia (nervosa)
Ribes aureum
Ribes nevadensis
Amelanchier florida
Cercocarpus (macrourus)
Crataegus (columbiana)
Heteromeles arbutifolia
Prunus demissa
Prunus emarginata
Rosa nutkana
Sorbus (californica)

Salix exigua	*Amorpha (californica)*
Salix hookeriana	*Acer glabrum*
Salix laevigata	*Acer macrophyllum*
Salix lemmonii	*Acer negundo*
Salix melanopsis	*Fraxinus (oregona)*
Betula fontinalis	*Arbutus (menziesii)*
Chrysolepis chrysophylla	*Vaccinium parvifolium*
Quercus chrysolepis	*Symphoricarpos oreophilus*

The quantitative data indicate that, apart from the subdominant species whose distribution was controlled chiefly by high water table, most of the fossil species contributed to the bordering mixed conifer-deciduous hardwood forest and sclerophyll forest (or woodland). Some of the allied living species, such as *Abies magnifica, Picea breweriana* and *Pinus balfouriana*, now occur chiefly at high elevations and contribute to fir and subalpine forests. However, there is no compelling evidence to support the general opinion that they require high relief in the Buffalo Canyon area. Geologic evidence shows that relief was generally low to moderate. Fossils representing these species occur in other floras in western Nevada and Oregon that also occupied areas of low relief, and thus they appear to have been integral members of Miocene mixed conifer-hardwood forests. Furthermore, there is no evidence that remains of montane conifers (winged seeds, foliage, etc.) could be transported from mountain sites 8-10 miles distant from the area of accumulation without destruction, and also enter the record in considerable abundance, as have the seeds of hemlock, spruce, and other montane taxa at Buffalo Canyon (Spicer and Wolfe, 1987; Axelrod, 1988). It was the later changes in climate, chiefly decreased precipitation and lowered equability, that restricted the modern species to higher elevations with cooler climate and a lower evaporation rate, as compared with the warmer climate and higher evaporation rate at lower levels today (see Axelrod, 1976a, fig. 4). That the modern, upland species occur in an area with some summer moisture may well account in part for their relict survival.

Many of these species in the Klamath Mountains range northward into the Cascades, and most also occur in the northern to central Rocky Mountains and border areas, where there are several additional species with similar (or allied) species in the fossil flora:

Abies (amabilis)	*Betula subcordata*
Larix occidentalis	*Mahonia aquifolium*
Picea (engelmannii)	*Ribes aureum*
Pinus ponderosa	*Ribes americanum*
Pseudotsuga menziesii	*Ribes (viscocissimum)*
Tsuga mertensiana	*Amelanchier florida*
Juniperus (scopulorum)	*Crataegus chrysocarpa*
Populus angustifolia	*Prunus demissa*
Populus (hastata)	*Prunus emarginata*
Populus tremuloides	*Rosa nutkana*
Salix exigua	*Sorbus (scopulina)*
Salix melanopsis	*Acer glabrum*
Betula fontinalis	*Acer negundo*
Betula (occidentalis)	*Vaccinium (occidentale)*
	Symphoricarpos (rotundifolius)

Species that do not now occur in California include *Larix occidentalis*, which prefers moist sites in the mixed conifer forest from northeastern Oregon north and east to central Idaho. *Betula subcordata* is allied to *B. cordifolia* of the eastern United States, and *B. occidentalis* is a western segregate of *B. papyrifera*. Vegetation in the vicinity of Priest Lake, central Idaho, and Kachess and Keechelus lakes in the central Cascades of Washington, includes many of the species listed above. Most of these taxa occur also in central Idaho where *Larix* has its southern limit. In this area about 150 mm (6 in.) of rain occurs during the warm season, and it increases northward to fully 250-305 mm (10-12 in.) along the western slopes of the Continental Divide in northern Idaho.

Several Buffalo Canyon species are allied to modern species that are now in warmer, chiefly forest-border areas, or contribute largely to broadleaved sclerophyll vegetation. Two of these, *Arbutus arizonica* and *Juglans major*, live in the mountains of southern Arizona and adjacent New Mexico and range southward into Mexico. In the Chiricahua Mountains they occur from the evergreen sclerophyll forest up into the lower mixed conifer forest. Among the species in this region allied or closely related to species in the fossil flora are the following:

Abies concolor	*Ribes* sp.
Pinus ponderosa	*Amelanchier* sp.
Pseudotsuga glauca	*Cercocarpus* sp.
Juniperus sp.	*Prunus melanocarpa*
Populus angustifolia	*Rosa* sp.
Populus tremuloides	*Acer grandidentatum*
Salix exigua	*Acer negundo*
Salix laevigata	*Arbutus (arizonica)*
Salix melanopsis	*Fraxinus velutina*
Juglans major	*Symphoricarpos rotundifolius*
Quercus chrysolepis	

In addition, sclerophyll woodland with *Arbutus* and *Juglans* ranges into Sonora and Chihuahua, Mexico, where *Populus brandegeei*, allied to the fossil *P. cedrusensis*, has a riparian occurrence. These taxa now in the southwestern United States and adjacent Mexico are in a region with ample summer as well as winter precipitation. *Arbutus xalapensis* occurs widely in the mountains of Mexico where it is represented by several varieties that show relationship to *A. trainii*. Two other taxa in the fossil flora, *Eugenia* and *Cedrela*, now occur farther south in Mexico. They represent relict occurrences at Buffalo Canyon.

Three additional sclerophyll species allied to the fossil taxa are found on the Channel Islands, southern California. *Cercocarpus blancheae* occurs on the northern Channel Islands and also at scattered sites on the summit of the Santa Monica Mountains bordering the coast. *Lyonothamnus asplenifolius* is allied to the fossil *L. parvifolius*, an extinct small-leaved species. *Mahonia insularis* of Santa Cruz Island, similar to the fossil *M. reticulata*, has been considered a subspecies of *M. pinnata* (Munz and Keck, 1959:109); further study may show that it is more nearly related to the Mexican species *M. gracilis* (Hart.) Fedde and *M. chochoco* (Schl.) Fedde.

A number of species in the fossil flora have allies in the eastern United States, including:

Tsuga (caroliniana)
Juniperus virginiana
Populus (balsamifera)
Populus tremuloides
Salix nigra
Alnus maritima
Betula cordifolia
Betula lenta
Betula papyrifera
Carpinus caroliniana
Carya ovata
Ulmus rubra

Hydrangea sp.
Ribes americanum
Amelanchier sp.
Crataegus spp.
Prunus virginiana
Rosa spp.
Sorbus americana
Amorpha angustifolia
Robinia pseudoacacia
Acer negundo
Acer saccharum
Fraxinus sp.
Vaccinium arbireum

Three of these genera (*Carpinus, Carya, Ulmus*) are no longer in the western United States. In addition, several species in the eastern United States, notably *Amorpha angustifolia, Betula cordifolia, B. lenta,* and *B. papyrifera,* are similar to Buffalo Canyon taxa. The other eastern species are allied to western species (see above), but the latter are more closely related to the fossils.

Finally, some modern species related to Buffalo Canyon taxa now occur only in east Asian temperate forests; the species in parentheses are allied to, but not the closest analogues of, the fossils:

Abies sp.
Picea (likiangensis)
Picea polita
Pseudotsuga sp.
Tsuga sp.
Juniperus spp.
Populus (szechuanica)
Populus (tremula)
Salix sp.
Alnus japonica
Betula (japonica)

Carya (chinensis)
Carpinus sp.
Zelkova serrata
Hydrangea aspera
Ribes spp.
Crataegus spp.
Prunus (wilsonii)
Prunus davidiana
Sorbus pohuashanensis
Acer (henryii)
Cedrela sinensis
Fraxinus chinensis

Of these, only *Zelkova* is restricted to Asia, distributed from Japan and China discontinuously westward to the Caucasus Mountains. Others that do not have close allies in North America include *Picea polita, Hydrangea aspera, Prunus davidiana,* and *Sorbus pohuashanensis.* These occupy areas of ample summer rain and contribute to mixed conifer-hardwood and bordering conifer forests.

To summarize, the fossil flora of 69 species includes taxa whose modern descendants, or their close allies, contribute to western mixed-conifer forests and to the broadleaved sclerophyll vegetation associated with them at lower elevations, and to the conifer-deciduous hardwood forests of the eastern United States and of eastern Asia. The fossil species allied to those now in these disjunct regions contributed to a richer western Miocene forest than any that has survived. The richest living segregate of the Miocene forest occurs in the Klamath Mountain region of northwestern California, owing to favorable climate and terrain.

PALEOECOLOGY

VEGETATION

The distributional and ecologic evidence outlined in the preceding summary forms a basis for reconstructing the probable plant associations in Buffalo Basin. However, inferences drawn from this evidence must be tempered by the relative abundance of the specimens, which indicates proximity to the site, though not to their dominance or rarity in the regional flora. Furthermore, in reconstructing Miocene vegetation, it is evident that fossil floras have been segregated into derivative communities that now occupy different subclimatic regions, so this must also be taken into account. The following plant associations appear to have characterized the basin.

Lake-shore Deciduous Forest

This community was composed chiefly of deciduous hardwoods and shrubs along the lake shore. To judge from the abundance of the specimens, these species probably contributed to a dominantly deciduous community. Since most of them occupied stream margins, seeps, and other moist sites in nearby valleys, they are also listed as members of the mixed conifer-deciduous hardwood forest.

Trees

Populus eotremuloides
Populus payettensis
Populus pliotremuloides
Alnus latahensis
Betula desatoyana

Betula thor
Acer negundoides
Fraxinus desatoyana
Fraxinus eastgatensis

Shrubs

Salix churchillensis
Salix laevigatoides
Salix owyheeana
Salix pelviga
Salix storeyana

Ribes barrowsae
Amelanchier desatoyana
Crataegus middlegatei
Prunus chaneyii
Prunus moragensis
Sorbus cassiana

Conifer-Hardwood Forest

This forest evidently covered slopes adjacent to the lake, reaching down to it in cooler sites along stream margins, where it probably mingled with plants of the lake shore deciduous forest.

TREES

Conifers

Abies concoloroides
Abies laticarpus
Larix churchillensis
Picea lahontense
Picea magna

Picea sonomensis
Pinus balfouroides
Pinus ponderosoides
Pseudotsuga sonomensis
Tsuga mertensioides
Chamaecyparis cordillerae

Hardwoods

Acer medianum
Acer negundoides
Acer oregonianum
Acer tyrrellii
*Betula desatoyana
*Betula idahoensis
*Betula thor
Carpinus oregonensis

Carya bendirei
Chrysolepis sonomensis
Fraxinus eastgatensis
Populus eotremuloides
Populus pliotremuloides
Robinia bisonensis
Ulmus speciosa
Zelkova brownii

SHRUBS
(Chiefly riparian or in moist sites in the forest)

Salix laevigatoides
Salix desatoyana
Salix owyheeana
Salix storeyana
Alnus latahensis
Betula ashleyii
Mahonia macginitiei
Ribes bonhamii
Ribes barrowsii

Ribes stanfordianum
Hydrangea bendirei
Amelanchier desatoyana
Prunus chaneyii
Prunus moragensis
Prunus treasheri
Rosa harneyana
Sorbus cassiana
Acer trainii
Vaccinium sophoroides

Sclerophyll Woodland

Exposed warm slopes probably supported plants that attained optimum development in the nearby region at lower elevations where climate was warmer. At Buffalo Canyon they probably were primarily confined to exposed, drier, well-drained south- and west-facing slopes provided by the hard, welded tuffs and flows that rimmed the area.

Trees	Shrubs
Arbutus trainii	*Amorpha stenophylla*
Cedrela trainii	*Cercocarpus ovatifolius*
Eugenia nevadensis	*Chamaebatia nevadensis*
Juniperus desatoyana	*Lyonothamnus parvifolius*
Quercus hannibalii	*Heteromeles desatoyana*
Quercus wislizenoides	*Ribes webbii*
	Symphoricarpos wassukana

Along stream margins in this zone were several taxa whose modern allies prefer warmer sites, notably *Fraxinus desatoyana, Juglans desatoyana, Populus cedrusensis, Salix churchillensis,* and *Salix pelviga.*

Summary

A Miocene aerial view looking north down Buffalo Basin would show that it was surrounded by low volcanic hills, except some 5 miles north of the fossil locality where there was somewhat higher relief in the ancestral Desatoya Range. The lake was deepest in its southeast corner near the fossil site and shallowed to the north, where it was impounded by quartz latite ashflow tuffs of the Desatoya Formation. There was some external drainage for the lake, which was in a collapse basin. The lake shore area and entering streams supported a dense forest of deciduous hardwoods, mainly species of *Acer, Betula, Fraxinus, Populus,* and *Salix,* as well as diverse shrubs, such as *Amelanchier, Crataegus, Mahonia, Prunus, Ribes, Rosa,* and *Vaccinium.* Bordering slopes were covered with a rich mixed conifer-deciduous hardwood forest with species of *Abies, Larix, Picea, Pinus,* and *Tsuga* associated with *Acer, Alnus, Betula, Carpinus, Robinia, Ulmus,* and *Zelkova,* together with mesic shrubs that included *Amelanchier, Crataegus, Mahonia, Prunus, Ribes,* and *Sorbus.* Drier, warmer south- and west-facing slopes on the hard welded tuffs and flows near the lake evidently supported an evergreen-sclerophyll woodland of *Arbutus, Cercocarpus, Chamaebatia, Eugenia, Juniperus, Lyonothamnus,* and *Quercus.* These taxa probably entered lower parts of the mixed conifer-hardwood forest in openings on warmer, drier slopes, much as their allies do today.

CLIMATE

The climatic requirements of modern species related to the fossils suggest that precipitation totaled about 890-1,000 mm (35-40 in.) annually. By contrast, it is near 380 mm (15 in.) at the locality today. Whereas precipitation now comes as winter rain and snow, and summers are dry, there was ample summer precipitation at Buffalo Canyon during the Miocene. This is indicated by the genera *Carpinus, Carya, Cedrela, Eugenia, Robinia, Ulmus,* and *Zelkova,* which are now confined to areas with warm-season rainfall. In addition, the fossil species of *Arbutus, Betula, Hydrangea, Juglans, Populus (cedrusensis), Prunus (treasheri),* and *Sorbus* are allied to modern species now in areas with ample summer rain.

Temperatures were more moderate than those now in this region. Today the mean annual range of temperature is 22.2°C (40°F). Summers are hot and winters are cold with many days below freezing, and snow is common. Milder temperatures were present in the Miocene because the Sierra Nevada, Carson Range, Virginia Range and Stillwater Mountains to the west were either very low or non-existent. This is clear, both from geologic evidence and from the composition of fossil floras in the area, whether from near the Sierran summit region (Niagara Creek, Ebbetts Pass, Carson Pass, Webber Lake, Mohawk) or in western Nevada (Pyramid, Sutro).

Inasmuch as many Buffalo Canyon species have modern allies on the western slope of the northern Sierra Nevada and in the Klamath Mountain region to the northwest, general thermal conditions in those areas most probably approximate temperatures of the Buffalo Canyon flora. However, in view of the trend to more extreme climates since the Miocene, it follows that the range of temperature ($\underline{A}$=amplitude) at Buffalo Canyon was less than that now in areas where there is mixed-conifer forest with broadleaved sclerophylls on nearby exposed warm slopes. The relations in Fig. 3 suggest that the mean annual temperature was approximately 10°C (50°F) and that the mean annual range of temperature was near 14°C (25°F). Using these estimates, and a normal sine wave for monthly temperatures, the mean conditions estimated for the flora are charted in Fig. 4. Note that the fossil flora is inferred to have no days warmer than ca. 17°C, but today the area has 4.4 months (132 days) warmer than that. Fig. 4 also indicates that the Buffalo Canyon flora had 3 months warmer than 15°C, but the nearby Eastgate flora had 4.8 months (144 days) warmer than that, consistent with the warmer environment of that area. Whereas no days at Buffalo Canyon were warmer than 17°C, the Eastgate flora had about 3.4 months (102 days) warmer than that. Today, 3.4 months (102 days) are warmer than 19.5°C.

Compared with conditions today (estimated in Chapter 2), frost frequency has increased from about 7% to 12%, or from 613 to 1051 hours of the year. Warmth has increased from $\underline{W}$ 12.5°C to $\underline{W}$ 13.6°C, or from 138 days warmer than 12.5°C to 171 days warmer than 13.6°C (for methodology, see Bailey, 1960, 1964, 1966; also, Axelrod and Bailey, 1976; Axelrod, 1981b). Equability or temperateness ($\underline{M}$=moderation) rating has decreased from near $\underline{M}$ 57 to $\underline{M}$ 48, reflecting the increasing range of post-Miocene temperature.

ELEVATION

Comparison of the inferred temperature of the fossil flora with that for the Early Miocene in the near-coastal area provides a basis for estimating elevation, because temperature decreases with elevation. The Valley Springs flora, dated at 22 Ma, comes from several sites in the lower foothills of the Sierra Nevada. The small collections, secured during the 1880s, were recovered from tunnels made in the rhyolitic Valley Springs Formation to get access to the Auriferous Gravels. The localities are at relatively low elevations: Chili Gulch, 395 m (1,300 ft.), Valley Springs, 213 m (700 ft.), and Wheats, 198 m (650 ft.). During the Miocene the area was even lower, for the Sierran block has since been elevated many thousands of feet (see Axelrod, 1980a, fig. 9). At a minimum, elevation of the Valley Springs Formation has increased on the order of one-third, which suggests that the fossil sites were then near 120 m (400 ft.) for Chili Gulch and 60 m (200 ft.) for Wheats and Valley Springs. The combined flora —

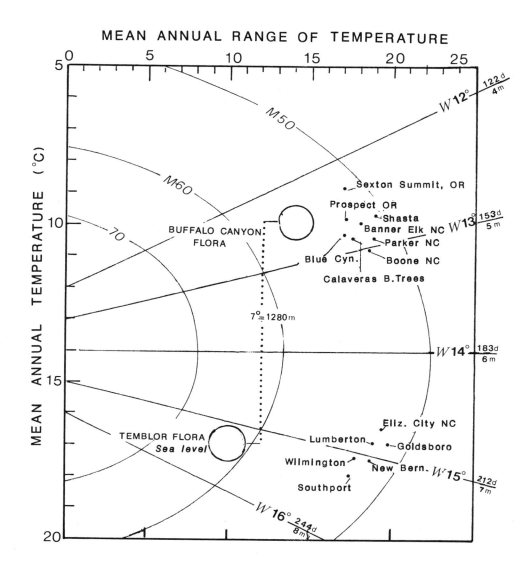

FIGURE 3. Temperature in areas where species show relationship to those in the Buffalo Canyon and Temblor floras, the latter at sea level. The difference in mean annual temperature provides a basis for estimating elevation of the Buffalo Canyon flora. The nomogram by Bailey (1960, 1964) shows warmth of climate (radiating lines) in terms of the number of days (or months) with mean temperature warmer than the stipulated condition. The arcs provide a measure of equability or temperateness (Bailey, 1964).

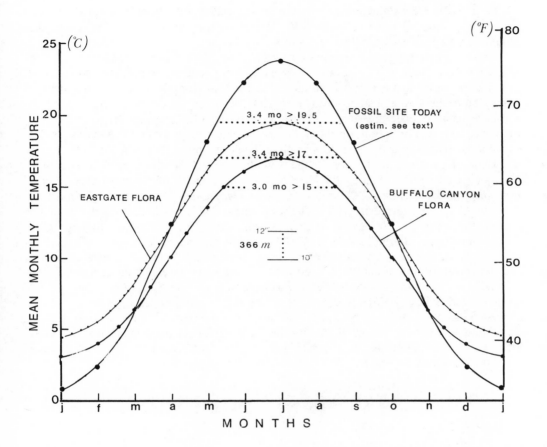

FIGURE 4. Estimated mean monthly temperatures for the Buffalo Canyon flora, compared with those at the site today (estimated from Eastgate), and that for the Miocene Eastgate flora. The Buffalo Canyon flora had 3.0 months with mean temperature 15°C, whereas the Miocene Eastgate flora had 3.4 months with mean temperature 17°C. Today the fossil area has 3.4 months at 19.5°C. In winter, the Buffalo Canyon flora had a mean low monthly temperature of 3.0°C, whereas the Eastgate flora had a mean low of 4.5°C. At present the winter low at the fossil site is 0.5°C. The difference in estimated mean annual temperature of 2°C between the Eastgate and Buffalo Canyon floras suggests their separation in elevation by 366 m (1,200 ft.).

composed of a few, chiefly scrappy specimens, largely too incomplete to compare directly with other fossil species or with modern taxa — was listed earlier (Axelrod, 1944a, p. 218). Reexamination of the material suggests that the flora includes *Pinus* sp., *Betula* cf. *thor*, *Quercus* cf. *scudderi*, *Quercus* sp. (evergreen), *Cocculus* sp., *Liquidambar* sp., *Chrysophyllum* sp., and *Ceanothus* sp.

These species suggested warm temperate climate, with a mean annual temperature near 17-18°C. There was a low range of mean annual temperature, probably about 10°C as judged from its taxa and also from its position not far from the marine embayment in the region to the west and southwest (Reed, 1933, fig. 31). A low range of temperature and an equable climate are suggested also by a cycad (cf. *Ceratozamia*) in the Lower Miocene Olcese Formation in the foothills of the southern Sierra Nevada (UCD collection). The Valley Springs flora is slightly older than the Buffalo Canyon, whereas the Temblor flora, from the base of the Temblor Formation near Coalinga, is a little younger (ca. 16 Ma). Among the Temblor taxa are species of *Glyptostrobus, Pinus, Bumelia, Carya, Castanea, Cedrela, Diospyros, Magnolia, Nyssa, Persea, Platanus, Quercus* (5 sp., 3 evergreen), and *Zelkova* (Renney, 1972). Allied modern vegetation occurs along the Atlantic Coast, from South Carolina into North Carolina. In that region mean annual temperature (T) is from 17-18°C, and the mean annual range (A) is also 17-18°C. The data suggest a difference in mean temperature between the Buffalo Canyon and the coastal floras of about 7°C and hence a difference in elevation of about 1,280 m (4,200 ft.), assuming a normal terrestrial lapse rate of -.55°C/100 m or 183 m/-°C (Axelrod and Bailey, 1976). Inasmuch as the nearby Middlegate-Eastgate floras had an estimated elevation near 913 m (3,000 ft.), the difference between them and the Buffalo Canyon flora was on the order of 367 m (1,200 ft.). The data also suggest that the fossil locality, now at 1,848 m (6,060 ft.), has been elevated on the order of 580 m (1,900 ft.) since the flora was living.

AGE

The Buffalo Canyon Formation rests unconformably on a thick sequence of volcanic rocks that range in age from 32 to 22 Ma (Barrows, 1971). At the mouth of Buffalo Canyon, and at two other sites in the Ashflow member, the welded tuffs have yielded an average K/Ar age of 18 Ma (see Appendix). This Ashflow member grades up into the Diatomite member in which the flora is about 80 m higher stratigraphically. Assuming a constant deposition rate, the flora is probably not much younger than 18 Ma. This is equivalent to the late Hemingfordian land-mammal Age and to the middle Saucesian foraminiferal Stage (see Turner, 1970).

The flora differs considerably from the Middlegate and Eastgate floras 22 and 17 km northwest, and separated from the Buffalo Canyon area by the Eastgate Hills (Axelrod, 1985). Those floras have a greater representation of broadleaved sclerophyll trees and shrubs and a poorer development of exotic genera and species. Whereas each of the exotics in those floras is represented by only 1 to 3 specimens, at Buffalo Canyon they are more abundant and some were co-dominants of the mixed conifer-deciduous hardwood forest.

The floristic differences between the floras are not due to a difference in age. The rise in elevation of ca. 366 m to the Buffalo Canyon site was probably responsible for much of the difference. This would result in higher precipitation and lower temperature, consistent with the composition of the Buffalo Canyon flora. In addition, its position on the windward slopes of the ancestral Desatoya Range would be expected to provide more summer precipitation, and hence a better representation of exotics, some of which (*Betula, Carya, Juglans, Ulmus, Zelkova*) are represented by numerous specimens.

REGIONAL RELATIONS

As more Miocene floras have been collected, it has become apparent that vegetation at any one time in the western United States was quite varied. This disposes of the earlier notion that vegetation showed relatively little change across broad regions, which led to the belief that the age of a flora could be determined by the number of species it had in common with one whose age was established. If there were few taxa in common with another flora, it presumably meant that they were of very different age.

It has been apparent for some time that contemporaneous floras at different elevations, or with different exposures (e.g., north-south vs. east-west slopes), may include taxa representing very different communities. The Eastgate and Middlegate floras from the same basin, situated scarcely 8 km (5 mi.) apart and at the same stratigraphic level, show that contemporaneous floras may differ considerably depending on exposure (Axelrod, 1985). Whereas the Middlegate flora derived its taxa from south-facing slopes, the Eastgate flora represents vegetation derived from chiefly north-facing slopes. The Creede flora (Axelrod, 1987) provides further evidence of the importance of local relief in determining composition. Florules from Birdsey Gulch, situated at the mouth of a major canyon with cold air drainage, represent a dominant conifer forest. The Dry Gulch florules, in a sheltered part of the basin 2.3 km east, represent a dominant juniper-piñon woodland, and forest taxa are subordinate.

Greater differences become apparent when floristic provinces are compared (Fig. 5). The Middlegate-Eastgate floras (Axelrod, 1985) show no relationship to the Tehachapi flora (17.5 Ma) from the western border of the Mohave Desert (Axelrod, 1939). In that semi-arid area, vegetation was dominated by evergreen sclerophyllous oaks, palm, avocado, evergreen shrubs, and drought-deciduous taxa representing thorn forest. Further, the floras of coastal California (Carmel, Puente, Topanga; UC Mus. Pal. collection) represent mesic oak-laurel-palm forests in which evergreen oaks dominated. Situated west of the San Andreas rift system, they were transported tectonically some 250-300 km north from their site of growth to their present area.

In the interior, from central Nevada to eastern Washington, floras like the Pyramid (UC Mus. Pal.), 49-Camp (LaMotte, 1936), Succor Creek, Mascall (Chaney, 1959), Spokane and Coeur d'Alene (Knowlton, 1926; Brown, 1937), and Grand Coulee (Berry, 1931) represent relatively lowland forests. They have mesic conifers (*Metasequoia, Taxodium, Glyptostrobus*) and abundant, exotic, deciduous hardwoods whose nearest relatives are now in the eastern United States and eastern Asia. Some of these floras

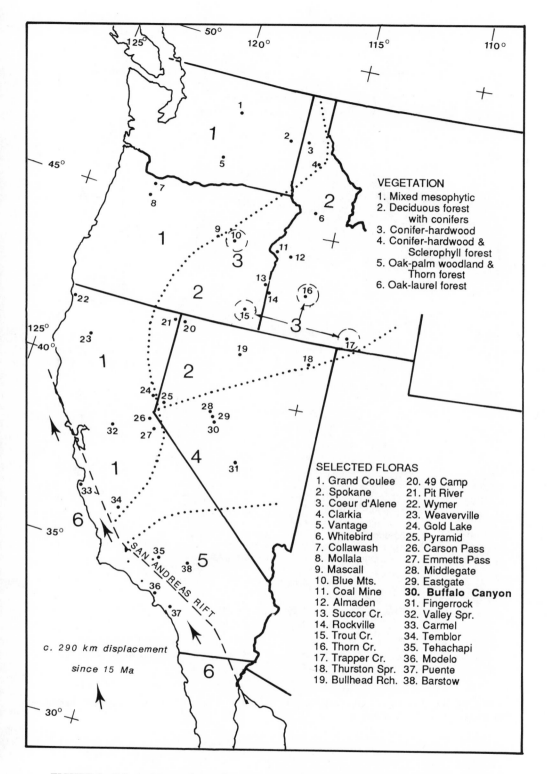

FIGURE 5. Principal vegetation regions during the Middle Miocene. Only the representative floras in each province are shown in this figure.

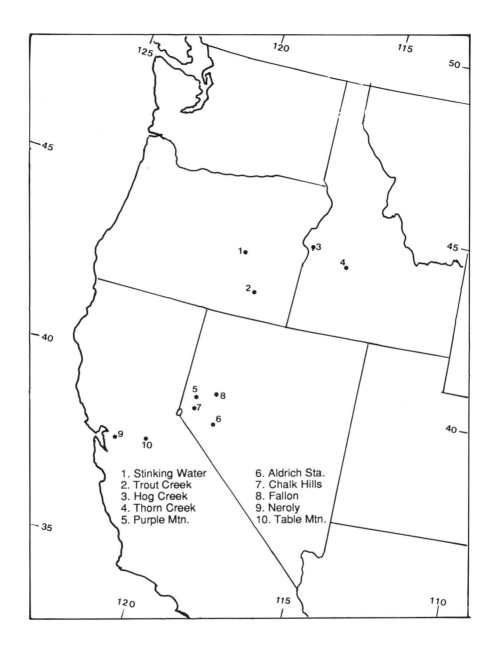

1. Stinking Water
2. Trout Creek
3. Hog Creek
4. Thorn Creek
5. Purple Mtn.

6. Aldrich Sta.
7. Chalk Hills
8. Fallon
9. Neroly
10. Table Mtn.

FIGURE 6.   Late Middle Miocene floras of generally similar age show marked changes in composition over moderate distances.

from comparatively lower elevations (i.e., Spokane, Clarkia; Smiley and Rember, 1958), also have evergreen species of Lauraceae and Magnoliaceae, implying an equable climate. With a rise in elevation in the interior region, montane conifers increase, warm temperate evergreens disappear, and the hardwoods are those of generally cooler requirements. Good examples are provided by the Blue Mountains, Trapper Creek, and Sparta floras. Regional differences in vegetation during the Middle Miocene are further accentuated by the rich floras from the upper Sierran slope, now near treeline, as shown by the Ebbetts Pass, Carson Pass, Webber Lake, and Gold Lake floras. They include more numerous mesic deciduous hardwoods (i.e., *Liquidambar, Platanus*) and warm temperate evergreen genera (i.e., *Magnolia, Nectandra, Persea, Quercus*) than floras of the same age in western Nevada 50-100 km (30-60 mi.) east. They find greater affinity with floras on the western slopes of the Miocene Oregon Cascades (lists in Peck et al., 1964), a moister region of high equability and precipitation.

The general boundaries between these Middle Miocene (19-15 Ma) floristic provinces persisted into the later Miocene (14-12 Ma) in the same areas. However, the younger floras in each province were less diverse in composition and had fewer exotic genera or species. These contrasting floras (see Axelrod, 1985, table 13) also demonstrate that vegetation in each region varied considerably over short distances (Fig. 6). This is well exemplified by the Purple Mountain (Axelrod, 1976b) as compared with the Aldrich Station flora (Axelrod, 1956); the Fallon (Axelrod, 1956) and Chalk Hills (Axelrod, 1962) floras of western Nevada; the Table Mountain (Condit, 1944) and Neroly (Condit, 1938) floras of west-central California; the Stinking Water (Chaney, 1959) compared with the Trout Creek (MacGinitie, 1933, Graham, 1965) of southeastern Oregon; and the Hog Creek (Dorf, 1936; Smith, 1938) and Thorn Creek (Smith, 1941) floras of southern Idaho.

In summary, the evidence shows the rapid response of Middle and Late Middle Miocene communities to changes in physical conditions over comparatively short distances. These younger floras all differ from slightly older ones in that they more nearly resemble modern western forests in composition. The change reflects the rapid decrease in summer rain that commenced over the interior region in the transition from 15 to 14 Ma. This is shown by two nearby Nevada floras, Pyramid (15.6 Ma) and Purple Mountain (14.8 Ma). Whereas the Pyramid flora is dominated (80+%) by exotic deciduous hardwoods of eastern affinity, they are largely absent from the overlying Purple Mountain flora (Axelrod, 1990). As outlined earlier, the marked change in composition appears to reflect the effect of the spreading Antarctic ice sheet on global climate (Axelrod, 1985).

SYSTEMATIC DESCRIPTIONS

For the most part, Buffalo Canyon leaves and seeds cannot be readily differentiated from those of living species. Certainly, if one searches for the minutest differences in the ultimate leaf venation or in the nature of marginal teeth compared with a similar modern species, differences may be found. Whether these are significant at the specific level must, for the most part, remain uncertain. The nature of finer nervation, or of leaf serration, may well vary with position on a tree (sun vs. shade or north vs. south exposure) over the range of a species. For example, there are significant differences in the size of samaras of *Acer macrophyllum* over its area, and the allied fossil species *A. oregonianum* Knowlton displayed similar differences in the Miocene (see page 63). Rather than attempt to detail minute differences in the fossil leaves or seeds as compared with those of similar modern species, it has seemed best to adopt a conservative stand and not describe differences that may not be significant at the species level. Although the fossils are very similar to modern species, we emphasize their separation in time from modern taxa by giving them separate names.

Family PINACEAE

Abies concoloroides Brown
(Plate 3, fig. 1)

Abies concoloroides Brown, Wash. Acad. Sci. Jour., 30, p. 347, 1940.
　　Axelrod, Univ. Calif. Pub. Geol. Sci. 33, p. 275, pl. 4, figs. 5, 6 only; pl. 12, fig. 6, 1956.
　　Chaney and Axelrod, Carnegie Inst. Wash. Pub. 617, p. 138, pl. 11, fig. 4, 1959.
　　Axelrod, Univ. Calif. Pub. Geol. Sci. 39, pl. 227, pl. 42, figs. 1-3, 1962.
　　Axelrod, Univ. Calif. Pub. Geol. Sci. 51, p. 105, pl. 5, figs. 4-10; pl. 7, fig. 1, 1964.
　　Axelrod, Univ. Calif. Pub. Geol. Sci. 129, p. 107, pl. 16, figs. 1-4, 1985.

Two wedge-shaped winged seeds with the wings attached to the upper part of the seed are in the flora. They differ from those of *A. laticarpus* MacGinitie, in which the wings are strongly asymmetrical and attached well down on the seed.

Seeds of *A. concoloroides* resemble those of *A. concolor* Lindley and Gordon, a widely distributed tree in the western United States. In California (dry summers), this fir is confined to the middle and upper part of the mixed conifer forest. However, in the summer-rain area of the Rocky Mountains and in the eastern and southern Great Basin, *A. concolor* descends into the lower mixed conifer forest and may be adjacent to piñon-juniper woodland.

Occurrence: Buffalo Canyon, hypotype 7920, homeotype 7921.

Abies laticarpus MacGinitie
(Plate 3, fig. 2)

Abies laticarpus MacGinitie, Carnegie Inst. Wash. Pub. 416, p. 47, pl. 3, fig. 5, 1933.
  Graham, Kent State Univ. Bull. Research Ser. 9, p. 57, 1965.
  Chaney and Axelrod, Carnegie Inst. Wash. Pub. 617, p. 139, pl. 11, fig. 8, 1959.
Abies concoloroides Axelrod, Univ. Calif. Pub. Geol. Sci. 33, p. 275, pl. 4, figs. 2-6;
  pl. 17, figs. 5, 6; pl. 25, fig. 5, 1956.
Abies concolor Wolfe, not Lindley, U.S. Geol. Surv. Prof. Paper 454-N, p. N4, pl.
  1, fig. 10; pl. 6, figs. 1-3, 6, 10, 11, 1964a.
Abies sp. Wolfe, ibid, p. N14, text fig. 5, 1964.

The winged seeds of this species are readily separated from those of *A. concoloroides* because the wings are strongly asymmetrical, much broader, and attached well down on the seed, not distally or nearly so.

The single specimen and its counterpart in the flora resemble the winged seeds of the red firs, *A. magnifica* Murray and especially *A. shastensis* Lemmon. This is suggested by the co-occurrence of similar seeds and cone scales with exserted bracts in the Miocene Alvord Creek and Purple Mountain floras. The former is found chiefly in the red fir forest, but the latter reaches well down into the mixed conifer forest, as in the Klamath Mountain region, on Mt. Shasta, and in the southern Sierra Nevada, where it is associated with *Sequoiadendron*.

Occurrence: Buffalo Canyon, hypotype 7922.

Larix churchillensis Axelrod, n. sp.
(Plate 5, figs. 2-5)

Larix nevadensis Axelrod, Univ. Calif. Pub. Geol. Sci. 129, p. 111, pl. 17, figs. 11-14, 1985.

The name *Larix nevadensis* Axelrod was used earlier for an Eocene larch from the Copper Basin flora (Axelrod, 1966). Represented by cone scales and needles, it evidently is a different species from the Eastgate larch and the present fossils. *L. churchillensis* (from Churchill County, Nevada) is therefore considered a new species represented by winged seeds in both the Eastgate and Buffalo Canyon floras.

Description: Winged seeds 1.2-1.5 cm long; wing acutely rounded distally, reaches under seed; seed ovate, 3-4 mm long, apex acute and acute proximally.

Discussion: Winged seeds of this larch are relatively abundant in the flora. The fine, filmy needles of larch have not been recovered, probably because they are quite fragile and would not readily withstand distant transport.

These seeds differ from those of *Tsuga*, which are larger and have a nearly rectangular outline. *Picea* seeds differ from those of larch in having an asymmetrical wing widest distally, and the tip of the seed is regularly snubbed.

These winged seeds are similar to those of *L. occidentalis* Nuttall, which occurs in the northern Rocky Mountains and southward into eastern Oregon. It prefers moist sites in the mixed conifer forest, and occurs in areas with summer rainfall totaling at least 150-200 mm (6-8 in.).

Occurrence: Eastgate, holotype 6708, paratypes 6709-11; Buffalo Canyon, hypotypes 7936-7939, homeotypes 7940-7941.

Picea lahontense MacGinitie
(Plate 3, figs. 7-10)

Picea lahontense MacGinitie, Carnegie Inst. Wash. Pub. 416, p. 46; pl. 3, figs. 6, 8
    (not fig. 4, which is *Picea sonomensis* Axelrod), 1933.
Pseudotsuga masoni MacGinitie, ibid., p. 47, pl. 3, figs. 1-3, 1933.
Picea magna MacGinitie. Axelrod, Univ. Calif. Pub. Geol. Sci. 33, p. 275, pl. 4, figs.
    7-12, 1956.
    Axelrod, Univ. Calif. Pub. Geol. Sci. 129, p. 112, pl. 4, figs, 8-10; pl. 17, figs.
    7-10, 1985 (see synonymy).

This realignment of taxa has been based in part on the occurrence in the Upper Bull Run flora (Eocene, 38 Ma, loc. 4) of abundant remains of *Picea* similar to those in the Trout Creek flora, including a cone as well as needles, winged seeds, and twigs. *P. lahontense* is one of at least 3 species of spruce in the upland Upper Bull Run flora, which is wholly dominated by montane conifers (Axelrod, 1968). In addition, similar remains, though not so abundant, are in the new collections of the Fingerrock flora, western Nevada.

The affinity of *P. lahontense* is with the larger coned species of eastern Asia, chiefly those in China, though the fossil species is certainly extinct.

Occurrence: Buffalo Canyon, hypotypes 7954-7956, 9347; homeotypes 7957, 7958, 7960-7965.

Picea magna MacGinitie
(Plate 4, figs. 10-13)

Picea magna MacGinitie, Carnegie Inst. Wash. Pub. 599, p. 83, pl. 18, figs. 5-7, 1953.
    Axelrod, Univ. Calif. Pub. Geol. Sci. 33, p. 275, pl. 4, figs. 7-12; pl. 25, figs. 8,
    9, 1956.
    Chaney and Axelrod, Carnegie Inst. Wash. Pub. 617, p. 140, pl. 12, figs. 10-15,
    1959.
    Axelrod, Univ. Calif. Pub. Geol. Sci. 51, p. 108, pl. 6, figs. 9-13, 1964.
    Wolfe, U.S. Geol. Surv. Prof. Paper 454-N, p. N15, pl. 1, figs. 3, 5; pl. 6, figs.
    7, 12, 17, 18, 22, 1964a.

Picea breweriana Wolfe, not S. Watson, ibid., p. N14, pl. 6, figs. 4, 5, 8, 9, 13 (not
  figs. 14 and 19, which are *Picea sonomensis* Axelrod), 1964a.

The winged seeds of this species are abundant in the flora, but no needles or twigs
of spruce are in the collection. This suggests that owing to their high area-weight ratio,
they were probably deposited near the lake shore.

These seeds, with short, stubby wings, show a general relationship to those of *P.
polita* (Siebold and Zuccarini) Carrière. In the absence of needles and cones that may
represent this species, it is best to assume that it is probably extinct.

Occurrence: Buffalo Canyon, hypotypes 7942-7945, 7948; homeotypes 7944, 7946,
7947, 7949-53, 7927, 7930-7935.

Picea sonomensis Axelrod
(Plate 3, figs. 13-16)

Picea sonomensis Axelrod, Carnegie Inst. Wash. Pub. 553, p. 190, pl. 36, fig. 2; p.
  251;pl. 42, figs. 2, 3, 1944.
Axelrod, Univ. Calif. Pub. Geol. Sci. 129, p. 113, pl. 4, figs. 6-7; pl. 17, figs. 3-6,
  1985 (see synonymy).

The winged seeds of this species are common in the Buffalo Canyon flora. They are
easily recognized by the small, snub-nosed seed attached to a wing that is conspicuously
rounded distally and much larger than the seed. These winged seeds are similar to those
produced by the weeping spruce, *P. breweriana* S. Watson, of northwestern California,
a rather rare tree that occurs in the upper mixed conifer forest and the fir and subalpine
forests above it. The fossil species had a wide distribution over the western states in the
Miocene and occurred also with broadleaved sclerophyll vegetation that occupied nearby
warmer, drier sites. Together with other presently upland species, the taxa were
confined gradually to higher elevations as summer rainfall decreased in the Late Miocene
and Pliocene.

Occurrence: Buffalo Canyon, hypotypes 7966-7969, homeotypes 7970-7980.

Pinus balfouroides Axelrod
(Plate 3, figs. 3-6, 17-19)

Pinus balfouroides Axelrod, Univ. Calif. Pub. Geol. Sci. 121, p. 209, 1980a (see
  synonymy and discussion).
Axelrod, Ann. Missouri Bot. Gard. 73, p. 613, figs. 22-33, 1986 (see synonymy
  and discussion).

The record of this pine in the Buffalo Canyon flora is based on several fascicles and
winged seeds. These can be separated from rather similar structures of *P. monticola*
Douglas because the latter has fascicles with smaller sheaths, so that the needles are
nearly deciduous. Also, its needles are slenderer, and the seed wings are not commonly
arched upward as are those of *P. balfouroides*.

The fossils closely resemble winged seeds and fascicles of *P. balfouriana* Greville
and Balfour, disjunct from the high Klamath Mountain region to the southern Sierra

Nevada. The fossil *P. balfouroides* is now known from several sites in western Nevada, which essentially links the modern populations in terms of their geologic history.

    Occurrence: Buffalo Canyon, hypotypes 7981, 7983-7986, 7997-7999; homeotypes 7982, 7987-7990, 7318-7319.

Pinus ponderosoides Axelrod
(Plate 3, figs. 11-12)

Pinus ponderosoides Axelrod, Ann. Missouri Bot. Gard. 73, figs. 63-67, 1986.
Pinus florissantii Axelrod, Univ. Calif. Pub. Geol. Sci. 33, p. 276, pl. 4, figs. 19-20;
    pl. 17, figs. 10-11, 1956.

    The record of this pine in the Buffalo Canyon flora is based on several seed wings that are dehiscent in the sense that the seed is not retained by the clasping wing.

    Among modern species, *P. ponderosa* Lawson produces similar seed wings. This is a common forest species found over much of the western United States.

    Occurrence: Buffalo Canyon, hypotypes 7991-7993, homeotypes 7994-7996.

Pseudotsuga sonomensis Dorf
(Plate 4, figs. 1-5)

Pseudotsuga sonomensis Dorf, Carnegie Inst. Wash. Pub. 412, p. 72, pl. 6, figs. 2-4,
    1930.
    Axelrod, Carnegie Inst. Wash. Pub. 553, p. 191, pl. 36, fig. 3; p. 251, pl. 42, fig.
    1, 1944
    Axelrod, Univ. Calif. Pub. Geol. Sci. 129, p. 116, pl. 4, fig. 16; pl. 16, figs. 15-
    19, 1985 (see other lit. citations).

    There are a number of winged seeds and several needles representing this species in the Buffalo Canyon flora. The winged seeds are relatively large, the wing is distinctly acute distally, and the seed is sharply truncated at an acute angle where it joins the wing. The needles are slender, have rounded apices, and are constricted at the base into a distinct petiole.

    The fossils are similar to needles and winged seeds of *P. menziesii* (Mirbel) Franco, which ranges widely in the western United States and reaches northward into the mountains of western Canada. The Rocky Mountain *P. glauca* (Beissner) Franco has smaller cones than the species farther west.

    Occurrence: Buffalo Canyon, hypotypes 9320-9324, homeotypes 9325-9332.

Tsuga mertensioides Axelrod
(Plate 4, figs. 6-9)

Tsuga mertensioides Axelrod, Univ. Calif. Pub. Geol. Sci. 51, p. 110, pl. 7, figs. 4-12,
    1964.
    Axelrod, Univ. Calif. Pub. Geol. Sci. 129, pl. 4, figs. 11-13; pl. 16, figs. 20-23,
    1985.

Winged seeds of this hemlock are relatively common in the flora, though no needles or cones were encountered in the collection. This suggests that the tree lived chiefly in the mixed conifer forest in the bordering hills, well removed from the area of plant accumulation.

These winged seeds are quite distinctive, for they are relatively large and the wing is essentially rectangular in outline and normal to the seed where it is attached to it. The fossils are similar to the winged seeds of *T. mertensiana* (Bongard) Carrière, a common montane forest tree from the Sierra Nevada of California northward to southwestern Alaska and eastward into the northern Rocky Mountains in Idaho and Montana. Although this hemlock is a high montane species in California and Oregon and bordering areas, the fossil is recorded with taxa representing broadleaved sclerophyll vegetation. As with other taxa of similar occurrence, the fossil species may represent an ecotype that ranged to lower elevations.

Occurrence: Buffalo Canyon, hypotypes 9333-9337; homeotypes 9336, 9338-9346, 7923-7926, 7928, 7929.

Family CUPRESSACEAE

Chamaecyparis cordillerae Edwards and Schorn
(Plate 4, fig. 18)

Chamaecyparis cordillerae Edwards and Schorn. Axelrod, Univ. Calif. Pub. Geol. Sci. 129, p. 118, pl. 4, fig. 21; pl. 17, figs. 21, 23; pl. 23, figs. 4-5, 1985 (see discussion and synonymy).

A single branchlet, temporarily "lost" among the numerous specimens of *Juniperus* in the flora, is certainly *Chamaecyparis*. The specimen is separable from *Juniperus* because the latter has ternate whorls on some shoots, and on other shoots displays a twisting of the axis or folding over of branches; and where ternate whorls or evidence of nonflattened foliage are lacking, specimens referable to *Juniperus* are more openly branched and more uniform in leaf morphology from branch-order to order than is *Chamaecyparis*.

The present material is more nearly allied to *C. lawsoniana* (A. Murr.) Parl. than to *C. nootkatensis* (Lamb) Spach, because the scale leaves are not so awl-like and all the scales have glands, in contrast to *C. nootkatensis*. *C. lawsoniana* occurs on the coastal slope of northwestern California and adjacent Oregon, and has a disjunct occurrence in the Mt. Shasta region on the Sacramento River.

Occurrence: Buffalo Canyon, hypotype 9892, 9892A.

Juniperus desatoyana Axelrod, n. sp.
(Plate 4, figs. 14-17)

Description: Slender, branching twigs from 4 to 5 cm long and 1.0-1.5 mm broad; clothed by small scale leaves, arranged in opposite pairs, those in facial view with a prominent gland, the tips of the scale leaves markedly acuminate-acute, the scale leaf reaching up to touch the scale above; small fruits, globose, 5-6 mm broad.

Discussion: Examination of a number of modern juniper species indicates that the fossils are most similar to those of the modern *J. occidentalis* Hooker. This is a common tree distributed along the Sierra-Cascade axis into southern Washington. It prefers drier sites for the most part, and at Buffalo Canyon its fossil ally most probably was on well-drained volcanic slopes provided by welded tuffs and andesite flows bordering the basin.

Occurrence: Buffalo Canyon, holotype 9355, paratypes 9349-9354, 9356-9358.

Family TYPHACEAE

Typha lesquereuxii Cockerell
(Plate 5, fig. 1)

Typha lesquereuxii Cockerell, Torrey Bot. Club 33, p. 307, 1906.
Typha latissima Lesquereux (not Al Braun), Rept. U.S. Geol. Surv. Terr. 8, p. 141, pl.
    23, figs. 4, 4a, 1883.

Leaves of cattail are common in the Buffalo Canyon shales. They are linear, ranging up to 2.0 cm broad, and some exceed 25 cm in length. They have numerous longitudinal veins interspersed with many fine linear veinlets crossed by short veins normal to the blade. There is no midrib in these specimens.

The fossils are similar to leaves of *T. latifolia* Linnaeus, which are typically broader than those of *T. angustifolia* Linnaeus. The presence of cattail in a lacustrine section is consistent with its preference for continually moist sites with a high water table.

Occurrence: Buffalo Canyon, hypotype 9358, homeotypes 9359-9365.

Family SALICACEAE

Populus cedrusensis Wolfe
(Plate 5, figs. 7-11)

Populus cedrusensis Wolfe, U.S. Geol. Surv. Prof. Paper 454-N, p. N16, pl. 7, figs.
    4, 5, 8; pl. 8, fig. 4, text-fig. 7, 1964.
    Axelrod, Univ. Calif. Pub. Geol. Sci. 129, p. 128, pl. 5, fig. 2; pl. 21, figs. 1-3,
    1985.
Populus sonorensis Axelrod, Univ. Calif. Pub. Geol. Sci. 33, p. 284, pl. 5, figs. 5, 9-
    11, 1956.

As noted previously, *P. cedrusensis* leaves in the Neogene floras of the central and southern Great Basin are commonly (or dominantly) elliptic, whereas those from the Mohave region and southern California are typically ovate. The presence of both sorts of leaves in the Buffalo Canyon flora, and in the large collection of the Stewart Spring flora, suggests an ecotonal relation between two closely allied populations. Since the great majority of leaves at Buffalo Canyon are elliptic, they are referred to *P. cedrusensis*. Judging from its associated species, *P. cedrusensis* lived in a cooler climate than did *P. sonorensis* in southern California and border areas.

Both species are allied to *P. brandegeei* Schneider (= *P. monticola* Brandegree) of the mountains of Baja California Sur and along streams in sclerophyll woodland vegetation at scattered sites in Sonora and Chihuahua.

Occurrence: Buffalo Canyon, hypotypes 9366-9369, 9377; homeotypes 9370-9376.

Populus eotremuloides Knowlton
(Plate 6, figs. 1, 3, 8)

Populus eotremuloides Knowlton, U.S. Geol. Surv. 18th Ann. Rept., pt. 3, p. 725, pl.
    100, figs. 1, 2; pl. 101, figs. 1, 2, 1898.
LaMotte, Carnegie Inst. Wash. Pub. 455, p. 114, pl. 5, figs. 7, 9, 1936.
Brown, U.S. Geol. Surv. Prof. Paper 186-J, p. 169, pl. 47, fig. 1, 1937.
Chaney and Axelrod, Carnegie Inst. Wash. Pub. 617, p. 151, pl. 17, fig. 4, 1959.
Axelrod, Univ. Calif. Pub. Geol. Sci. 129, p. 129, pl. 6, fig. 5; pl. 20, fig. 6; pl.
    21, fig. 6, 1985.

As discussed earlier (Axelrod, 1985, p. 129-130), fossil leaves of the *P. trichocarpa (hastata)* alliance are not always easy to separate when only one or two specimens are available. *P. trichocarpa* leaves tend to be more nearly ovate, whereas those of *P. balsamifera* are typically lanceolate with attenuated apices. In the Buffalo Canyon flora there are leaves typical of the *trichocarpa (hastata)* type, hence identified as *P. eotremuloides*. A few slenderer leaves are presumed to also represent that species rather than a *balsamifera*-allied species (i.e., *P. bonhamii* Axelrod), because leaves of the *trichocarpa (hastata)* alliance include both sorts, but *balsamifera-bonhamii* rarely does.

Occurrence: Buffalo Canyon, hypotypes 9378, 9379; homeotypes 9380, 9381.

Populus payettensis (Knowlton) Axelrod
(Plate 22, figs. 1-3)

Populus payettensis (Knowlton) Axelrod, Carnegie Inst. Wash. Pub. 553, p. 253, 1944b.
Rhus payettensis Knowlton, U.S. Geol. Surv. 18th Ann. Rept., pt. 3, p. 733, pl. 101,
    figs. 6, 7, 1898.
Populus payettensis (Knowlton) Axelrod, Univ. Calif. Pub. Geol. Sci. 129, p. 131, pl.
    5, figs. 1, 3; pl. 21, figs. 4-5; pl. 22, fig. 3, 1985 (see synonymy).

This species is represented by slender, lanceolate leaves. Most are nearly entire, though some have a few subdued glandular teeth. The petioles are somewhat longer than in most previously figured specimens, though they are readily matched by variation of the living *P. angustifolia* James. This is chiefly a Rocky Mountain cottonwood, ranging westward into the Great Basin. There is a relict population in the San Bernardino Mountains of southern California in the upper Santa Ana River drainage, an area with 65-75 mm summer rain.

Occurrence: Buffalo Canyon, hypotypes 9886-9888.

Populus pliotremuloides Axelrod
(Plate 5, fig. 6; Plate 7, figs. 9-11)

Populus pliotremuloides Axelrod, Carnegie Inst. Wash. Pub. 476, p. 169, pl. 4, figs. 1-
3, 1937.
Axelrod, Univ. Calif. Pub. Geol. Sci. 129, p. 132, pl. 20, figs. 2-3, 5, 1985 (see
citations and discussion).

There are 18 small, orbicular leaves that range from entire to those with crenate
margins in the flora. These are similar to leaves of *P. tremuloides* Michaux, one of the
most widely distributed trees in North America.

Leaves of *P. pliotremuloides* are consistently smaller than those of *P. voyana* Chaney
and Axelrod from the Miocene of the Columbia Plateau region (Chaney and Axelrod,
1959, p. 152, pl. 18, figs. 1, 3, 4). A similar species described from the Miocene of
Alaska, and designated *P. kenaiana* Wolfe (1966 and 1980), differs in no fundamental
way from *P. voyana*.

Occurrence: Buffalo Canyon, hypotypes 9382, 9386, 9388; homeotypes 9383-9385,
9387.

Salix churchillensis Axelrod, n. sp.
(Plate 6, figs. 4-7)

Description: Leaves linear, 2.5-6.0 cm long and 0.3-0.5 cm broad; tip acute, base
narrowly cuneate; petiole short, thick or thin; midrib firm; margin entire; alternate
secondaries diverge at medium to low angles, looping well up into blade to simulate a
marginal vein though mostly obscured, possibly because of thick pubescence as in
modern species; tertiaries not visible; texture firm.

Discussion: Leaves of this species are well represented in the collection. They
display close relationship with those of the modern sandbar willows, especially *S. exigua*
Nuttall, which is widely distributed in the western United States.

Other fossil species have been compared with the modern sandbar willows, but none
seem to represent this species. *S. payettensis* Axelrod (1944, p. 43, fig. 9) differs in its
serrate margin, and the tip is blunt. *S. edenensis* Axelrod (1937, pl. 4, fig. 7) is a much
smaller leaf and seems more nearly allied to *S. interior* Rowdee. Examination of the
specimens of *Salix* sp. figured by Dorf (1930, p. 8, fig. 3) indicates that they represent
Sapindus. The leaf of *S. taxifolioides* MacGinitie (1953, p. 23, fig. 2) is broader than
the present specimens and markedly toothed.

Occurrence: Buffalo Canyon, holotype 9389, paratypes 9390-9407.

Salix desatoyana Axelrod
(Plate 22, fig. 8)

Salix desatoyana Axelrod, Univ. Calif. Pub. Geol. Sci. 129, p. 133, pl. 22, figs. 2, 6,
7, 1985.

As noted earlier (Axelrod, 1985, p. 133), leaves in the Buffalo Canyon flora are
linear-lanceolate in shape, with numerous secondaries that loop to form a marginal vein,

and the margin has numerous sharp serrate teeth. This species is similar to leaves of the modern *S. nigra* Marshall. This is a widely distributed tree along rivers in the eastern United States.

An allied fossil species is *Salix truckeana* (Chaney, 1944, pl. 52, figs. 2-6), but it is broader and shows affinity with *S. gooddingii* Ball of the southwestern United States, reaching into central California.

Occurrence: Buffalo Canyon, holotype 6881; Eastgate, paratypes 6882, 6883.

Salix laevigatoides Axelrod
(Plate 8, figs. 3, 4)

Salix laevigatoides Axelrod, Carnegie Inst. Wash. Pub. 590, p. 55, pl. 2, fig. 10, 1950.
    Axelrod, Univ. Calif. Pub. Geol. Sci. 39, p. 231, pl. 46, fig. 4, 1962.

Several lanceolate leaves in the Buffalo Canyon flora are referred tentatively to this species which they resemble. The leaves are from 4.5 to 5.0 cm long with numerous subparallel looping secondaries, and the margin varies from entire to finely serrate. The fossils resemble leaves of *S. laevigata* Bebb and also those of several other living willows, notably *S. bebbiana* Sargent, *S. lutea* Nuttall, and *S. monticola* Bebb. Until a more complete suite becomes available, it seems best to refer the present material to a previously described species, which it closely resembles.

Occurrence: Buffalo Canyon, hypotypes 9408, 9409; homeotypes 9410-9413.

Salix owyheeana Chaney and Axelrod
(Plate 8, figs. 1, 2)

Salix owyheeana Chaney and Axelrod, Carnegie Inst. Wash. Pub. 617, p. 154, pl. 18,
    fig. 2, 1959.
    Axelrod, Univ. Calif. Pub. Geol. Sci. 39, p. 231, pl. 46, fig. 7 only, 1962.
Arbutus prexalapensis Axelrod, Univ. Calif. Pub. Geol. Sci. 33, pl. 32, fig. 3 only,
    1956.

Several elliptic-ovate leaves in the collection represent this species. They range in length from 5.5 to 7.0 cm and average 2.0-2.5 cm broad, though one large specimen (from a sucker shoot?) is 10 cm long and 3.5 cm broad. Apices are acute, the base obtuse or broadly acute. There are strong, widely spaced secondaries looping up along the margin, and prominent intersecondaries. The tertiaries are coarse, forming an irregular polygonal net and enclosing a similar coarse net of finer nervilles.

These specimens are similar to leaves of *S. hookeriana* Barratt, distributed chiefly in the coastal strip, from north-coastal California to British Columbia.

Occurrence: Buffalo Canyon, hypotypes 9414, 9915; homeotypes 9416-9418, 9889.

Salix pelviga Wolfe
(Plate 7, fig. 5; Plate 8, figs. 5, 6)

Salix pelviga Wolfe, U.S. Geol. Surv. Prof. Paper 454-N, p. N18, pl. 8, figs. 1, 2, 8,
    text-fig. 8, 1964.

Axelrod, Univ. Calif. Pub. Geol. Sci. 129, p. 134, pl. 6, figs. 6-7; pl. 22, figs. 4, 8, 12, 1985 (see synonymy).

Several long, slender leaves in the flora, ranging from sparsely toothed to essentially entire, with the teeth blunt and evidently gland tipped, represent this species. The fossils display the leaf variation of the modern *S. melanopsis* Nuttall, a shrub to small tree distributed from central California northward into British Columbia and east into the Rocky Mountains.

This species is also present in the Eastgate and Middlegate floras of the nearby Middlegate basin (Axelrod, 1985), as well as in the Stewart Spring flora to the south (Wolfe, 1964a).

Occurrence: Buffalo Canyon, hypotypes 9419, 9420; homeotypes 9421-9428, 9435.

Salix storeyana Axelrod
(Plate 6, fig. 2; Plate 7, figs. 1-4, 6)

Salix storeyana Axelrod, Univ. Calif. Pub. Geol. Sci. 121, p. 211, 1980.
    Axelrod, Univ. Calif. Pub. Geol. Sci. 39, p. 231, pl. 45, fig. 2; pl. 26, figs. 1-3, 5, 1962.
    Axelrod, Univ. Calif. Pub. Geol. Sci. 129, p. 135, pl. 6, figs. 3-4; pl. 23, figs. 7-8, 1985.

Numerous leaves in the flora are characterized by thin, looping subparallel secondaries that reach well up along the margin of the blade. Similar leaves are recorded in the Chalk Hills (Axelrod, 1962) and Middlegate and Eastgate floras (Axelrod, 1985). All of them resemble leaves of the modern *S. lemmonii* Bebb, which occurs in the mixed conifer and fir forests from the southern Cascades into southern California.

Occurrence: Buffalo Canyon, hypotypes 9429-9431, homeotypes 9432-9434, 9436-9439.

Salix sp., capsules of
(Plate 7, figs. 7, 8)

Several clusters of small capsules of willow are in the collection. Relationship with any one of the several described species in the flora is uncertain. The capsules are ovate, open at the tip, 2 mm broad near middle, 3-4 mm long, apex of each locule acute, short stipe 0.1 mm long.

Occurrence: Buffalo Canyon, nos. 9440-9442.

Family BETULACEAE

The problem of identifying fossil leaf species of alder and birch is not an easy one, because there is considerable overlap in leaf morphology of species of these genera, and no one character provides a reliable basis for discriminating between genera. The procedure followed here has relied chiefly on a comparison of suites of fossil leaves with the variation shown by modern species in the herbarium. Finer nervation is best seen in cleared leaves. However, since one or two cleared leaves cannot indicate the variation in leaf morphology of a species, it may lead to an overmultiplication of fossil species (see Axelrod, 1985, pp. 138-140).

Alnus latahensis Axelrod, n. sp.
(Plate 11, figs. 1, 2)

Prunus rustii Knowlton. Berry, U.S. Geol. Surv. Prof. Paper 154-H, p. 252, pl. 55, fig. 1, 1929.
Alnus relatus (Knowlton) Brown, U.S. Geol. Surv. Prof. Paper 186-J, p. 170, pl. 49, figs. 1-6, 1937.
Chaney and Axelrod, Carnegie Inst. Wash. Pub. 617, p. 159, pl. 22, figs. 1, 2, 6, 7, 1959.
Axelrod, Univ. Calif. Pub. Geol. Sci. 51, p. 116, pl. 9, figs. 6-11, 1964.

Description: Leaves elliptical; 5-10 cm long and 2-4 cm broad near middle; apex acuminate, base acute; midrib firm, straight; petiole thick, 2+ cm long; 6-7 pairs of secondaries diverging at medium angles and relatively straight in outward course, entering larger teeth; margin biserrate, with large and small teeth widely spaced; cross-tertiaries to secondaries are thin and irregular in course; 4th-order mesh irregularly polygonal; texture medium thick.

Discussion: Wolfe (1966, B16) pointed out that the specimen of *Phyllites relatus* Knowlton (1926, pl. 28, fig. 8), which was subsequently transferred to make *Alnus relatus* by Brown (1937, p. 170), is incomplete and is best considered a *nomen nudum*. To replace it, Wolfe has recognized *Alnus largei* (Knowlton) Wolfe and included in it specimens from other floras previously identified as *A. relatus*. However, Wolfe designated the second specimen of *B.? largei* Knowlton (1926, pl. 17, fig. 2) as the type of *A. largei* (Knowlton) Wolfe and included in it specimens previously identified as *Alnus relatus* (Knowlton) Brown. This is incorrect for two reasons. First, the type of *B.? largei*, clearly described by Knowlton (1926, pl. 17, fig. 1), is broadly ovate, not elliptic, in outline. Second, the specimen designated by Wolfe as *A. largei* (Knowlton) Wolfe (Knowlton's pl. 17, fig. 2) has strong tertiaries diverging from the basal side of secondaries, but not in the type specimen of *B.? largei* (pl. 17, fig. 1), nor in specimens of *A. relatus* (Knowlton) Brown. The type of *B.? largei* Knowlton (pl. 17, fig. 1) is similar to leaves of *A. rugosa* Sprengl, a large riparian shrub in the eastern United States.

In any event, the fossils from the Miocene of Oregon, Washington, and border areas previously identified as *A. relatus* need a new name. No fossils previously described as *Alnus* seem available. The types of *Prunus rustii* Knowlton from the Latah flora (Knowlton, 1926) may represent an alder allied to *A. rhombifolia* (Brown, 1937, p. 170). However, the specimen later identified by Berry (1929, pl. 55, fig. 1) as *P. rustii* is similar to those previously identified as *A. relatus*, all of which are herewith named *A. latahensis* Axelrod. The specimen of *P. rustii* Berry (1929, pl. 55, fig. 1) is designated as the type.

Three elliptic leaves in the Buffalo Canyon flora have widely spaced small teeth and are similar to those in Miocene floras to the north. These resemble leaves of *A. maritima* (Marshall) Muhlenberg of the southeastern states.

Similar foliar species have been described from the Miocene and Oligocene of Japan, notably *A. ezoensis* Tanai (1970) and others to which Tanai (1970, p. 466) refers.

Occurrence: Buffalo Canyon, hypotypes 9445-9446.

Alnus sp.
(Plate 11, figs. 3, 4)

Two clusters of cones in the flora are not now assigned to any one species. Those on Plate 11, fig. 4 are globose and much smaller than those normally produced by *A. latahensis*. Cones figured on Plate 11, fig. 3, are elliptic in outline and may represent a different species. However, variation of this sort occurs in white alder, *A. rhombifolia* Nuttall, though leaves of a species allied to it have not been recognized in the flora.

Occurrence: Buffalo Canyon, nos. 9444, 9447.

Betula ashleyii Axelrod
(Plate 10, figs. 6-11)

Betula ashleyii Axelrod, Univ. Calif. Pub. Geol. Sci. 51, p. 116; pl. 10, figs. 3-5, 1964.

Numerous specimens provide a basis for a better understanding of this species recorded earlier from the Trapper Creek flora, where it was based on a few specimens.

Supplementary description: Leaves variable in shape⁻lanceolate, ovate, to deltoid; blade 1.5-3.0 cm long, petiole 1-2 cm long; apex acute, base subacute, rounded or subcordate; 5-6 pairs of alternate secondaries diverging at moderate angles, straight or slightly curved above, reaching into marginal teeth or teeth supplied by strong tertiaries; tertiary mesh coarse, irregularly polygonal, some tending to a weak percurrency; finer intercostal veins form irregular polygons; margin singly or doubly serrate; teeth glandular; texture medium to thin.

Discussion: The leaves of this species are similar to those of *B. fontinalis* Sargent, a widely distributed large shrub or small tree, ranging from the central and southern Rocky Mountain region westward through the Basin and Ranges to the east front of the Sierra Nevada, and into the Klamath Mountain region in northern California.

Occurrence: Buffalo Canyon, hypotypes 9448, 9449, 9451, 9462, 9463, 9468; homeotypes 9450, 9452-9461, 9464-9467.

Betula desatoyana Axelrod, n. sp.
(Plate 10, figs. 1-5)

Description: Leaves broadly ovate to suborbicular, typical blades 4.5-5.0 cm long and 3.0-3.5 cm broad; apex acute, base cordate to subcordate; petiole 1.5-2.0 cm long, 1 mm broad, medium to thin; midrib straight, may become slightly curved above to somewhat wavering near apex; pairs of 5-6 secondaries, straight, diverging at about 45°; strong tertiaries trend directly into marginal teeth; interblade venation coarsely polygonal, with the finer 5th-order mesh open ended; margin biserrate in middle of blade, simpler above; texture medium.

Discussion: Leaves of birch dominate the Buffalo Canyon flora together with other typically stream- and lake-margin taxa. The variation in this large suite of specimens is not like that in any other collection known to me. The suite is quite different than that in the nearby Eastgate flora, where an allied birch, *B. thor* Knowlton, is reasonably abundant. The leaves of *B. desatoyana* are especially similar to those produced by *B. cordifolia* Regel of the northeastern states and adjacent Canada.

The large suite of broadly ovate cordate leaves readily separates this species from *B. thor*, which has typically ovate leaves. Although *B. cordifolia* has been considered a variety of *B. papyrifera*, the present record implies that both species have been distinct since the Middle Miocene.

Occurrence: Buffalo Canyon, holotype 9476; paratypes 9469-9475, 9477-9483.

Betula idahoensis Smith
(Plate 9, fig. 4)

Betula idahoensis Smith, Torrey Bot. Club Bull. 66, p. 470, pl. 10, fig. 4, 1939.

A single incomplete leaf in the flora appears to represent this birch. The specimen is long-ovate in outline, tapering above and has numerous parallel secondaries that curve distally at the margin. The margin is finely and evenly serrate, which gives the leaf an even profile.

Among living species, the specimen resembles leaves of *B. lenta* Linnaeus of the eastern United States. It differs from *B. vera* Brown (Brown, 1937), which has been compared with *B. lutea* Michaux, in that *B. vera* commonly has a cordate base and the leaf margin is not so evenly serrate. However, one of the specimens figured by Brown (1937, pl. 48, fig. 8) seems more nearly referable to *B. idahoensis* than to *B. vera*.

Occurrence: Buffalo Canyon, hypotype 9484.

Betula thor Knowlton
(Plate 9, figs. 1-3, 5, 7)

Betula thor Knowlton, U.S. Geol. Surv. Prof. Paper 140, p. 35, pl. 17, fig. 3, 1926.
    Chaney and Axelrod, Carnegie Inst. Wash. Pub. 617, p. 160, pl. 23, figs. 2-6, 1959
        (see synonymy).
    Axelrod, Univ. Calif. Pub. Geol. Sci. 129, p. 141; pl. 23, figs. 3, 6, 1985.

The leaves of this species, which are second in abundance in the flora, are similar to those of the eastern paper birch, *B. papyrifera* Marshall and its western segregate, *B. occidentalis* Nuttall, considered by some a variety of *B. papyrifera*. The latter occurs chiefly in the Pacific Northwest, adjacent Canada, and northern Rocky Mountains. It has been confused taxonomically with *B. fontinalis* Sargent, which occurs farther south. These birches are members of hydric communities that border streams and lakes where there is ample moisture throughout the year.

There is considerable variation in shape and margin of the fossil leaves, probably due in part to the large sample which includes relatively young leaves, as well as those from sucker shoots.

The large suite of *Betula* leaves in the Buffalo Canyon flora shows that *B. smithiana* (Axelrod) Axelrod from the Aldrich Station flora is a derived member of the *B. thor-papyrifera* complex. It is foreshadowed in the Buffalo Canyon flora by a few leaves (nos. 9896-9899) that are essentially identical with those in the Aldrich Station flora (Axelrod, 1956, pl. 6, figs. 5, 10, 15), dated at ca. 13-12 Ma. They differ from *B. thor* in having fewer marginal teeth and are more nearly related to *B. occidentalis* than to *B. fontinalis*. The Buffalo Canyon specimens appear to foreshadow the drier climate of the later Miocene. There are similar specimens in the Stewart Spring flora of subhumid requirements, judging from the large collection made by Howard Schorn.

Occurrence: Buffalo Canyon, hypotypes 9449, 9485-9486, 9488, 9490; homeotypes 9491-9493, 9488-9496, 9443-9445, 9896-9899, of *B. smithiana*, see text.

Family CARPINACEAE

Carpinus oregonensis Axelrod, new name
(Plate 9, fig. 6)

Carpinus grandis Unger. Chaney, Carnegie Inst. Wash. Pub. 346, p. 105, pl. 9, figs. 10, 11 (not figs. 7, 8, 9, which are *Engelhardtia*), 1927.

A single specimen in the Buffalo Canyon flora appears to represent *Carpinus*, and is well matched by leaves of *C. caroliniana* Walters. The leaf is ovate-elliptic, widest in the lower part, the tip is acuminate, the base broadly rounded, the essentially parallel secondaries do not have strongly developed abaxial tertiaries as in *Ostrya*, and the principal teeth are bordered basally by 2 or 3 small sharp teeth.

C. caroliniana is a common small tree in the deciduous forests of the eastern United States, reaching from the Gulf Coast northward through eastern Texas and Nebraska to the Great Lakes and southern New England.

C. fraterna Lesquereux (see MacGinitie, 1953) is an allied, slenderer-leaved species. Whether the bract described as *C. payettensis* Smith (1941) is to be associated with this species is not presently determinable.

The Bridge Creek specimens require a new name, for they do not represent the European species of a genus that had, and still has, many species. From an evolutionary standpoint, it is unlikely that one species of an active genus could have had a distribution from western Europe to Oregon and Nevada.

Occurrence: Bridge Creek, syntype UCMP 33, paratypes UCMP 32, 28; Buffalo Canyon, hypotype 9497.

Family JUGLANDACEAE

Carya bendirei (Lesq.) Chaney and Axelrod
(Plate 11, figs. 5-7)

Carya bendirei (Lesquereux) Chaney and Axelrod, Carnegie Inst. Wash. Pub. 617, p. 155, pl. 19, figs. 1-5, 1959 (see synonymy).

The numerous leaflets of hickory represented in the flora range from obovate to ovate to broadly elliptic in outline. They have a conspicuously serrate margin, and the numerous secondaries diverge at high angles and then loop upward to divide into tertiaries near the margin to supply the teeth.

Among modern species, the present material cannot readily be separated from leaflets of *C. ovata* (Miller) Koch of the eastern United States. This tree attains large size along stream banks and in low, moist sites. A similar adaptation probably accounts for its abundance in the flora.

Occurrence: Buffalo Canyon, hypotypes 9498-9501, homeotypes 9502-9506.

Juglans desatoyana Axelrod n. sp.
(Plate 13, figs. 1-4)

Description: Leaflets lanceolate, oblong-lanceolate, semi-falcate, from 4.0-7.0 cm long, 1.2-2.0 cm broad in widest part of blade, which is in middle to lower part; tips acute, base asymmetrically rounded to acute; petiolule short and thick, 4-6 mm long; 14-16 secondaries diverging at 35° to 45°, gently curving, supplying marginal teeth chiefly by branching tertiaries; principal tertiaries of intercostal area largely percurrent, enclosing quaternary polygonal areas in which the 5th-order venation is dense and forms an irregular network; margin with 2-3 teeth between secondaries, the teeth both acute and blunt; texture firm.

Discussion: This species is represented by numerous leaflets. They are similar to one of the leaflets identified by Wolfe (1964a, pl. 8, fig. 10 only) as *Juglans major* Torrey. The other specimen (fig. 9) is broader and of a sort that is not in the present suite; it may represent *J. alvordensis* Axelrod.

Leaflets of *J. desatoyana* are similar to the narrowly lanceolate leaflets of *J. major* Torrey in the American Southwest and northern Mexico, though some fossil leaflets tend to be more attenuated. The allied *J. rupestris* Englemann has slenderer, more falcate leaflets than the fossil. *J. major* is a regular riparian member of the sclerophyll woodland, reaching down into the piñon-juniper woodland belt, with scattered trees extending down drainageways well out into the grasslands below.

Occurrence: Buffalo Canyon, holotype 9507, paratypes 9508-9518.

Family FAGACEAE

Chrysolepis sonomensis Axelrod
(Plate 14, figs. 1, 2)

Chrysolepis sonomensis Axelrod, Univ. Calif. Pub. Geol. Sci. 129, p. 144, pl. 10, figs. 1-3, 1985 (see synonymy).

Several long-lanceolate leaves are similar to those produced by *Chrysolepis (Castanopsis) chrysophylla* (Douglas) Helmquist. This is a large spreading evergreen tree in the forests from the central Coast Ranges north into Oregon, and in the lower Cascade forests as well. A few isolated stands are in the central Sierra Nevada.

The leaves differ from the similar-shaped ones occasionally produced by *Q. chrysolepis* in having a more open 4th- and 5th-order venation.

Occurrence: Buffalo Canyon, hypotypes 9540, 9541; homeotypes 9542-9545.

Quercus hannibalii Dorf
(Plate 12, figs. 1-7)

Quercus hannibali Dorf, Carnegie Inst. Wash. Pub. 412, p. 86, pl. 8, fig. 9 only, 1930.
   Chaney and Axelrod, Carnegie Inst. Wash. Pub. 617, p. 168, pl. 24, fig. 2; pl. 25,
      figs. 11-13, 1959 (see synonymy).
   Axelrod, Univ. Calif. Pub. Geol. Sci. 121, p. 56, pl. 7, figs. 1-5, 1980a.
   Axelrod, Univ. Calif. Pub. Geol. Sci. 129, p. 146, pl. 8, figs. 1-7; pl. 25, figs. 1-7,
      1985.
Quercus chrysolepis sensu Wolfe, not Liebmann, U.S. Geol. Surv. Prof. Paper 454-N,
      p. N21, pl. 2, figs. 1-10, 14; pl. 9, figs. 2, 3, 5-7, 12, 16, 1964.

The leaves of this species form the most abundant specimens in the Buffalo Canyon
sediments. They are chiefly entire, though there are some serrate ones in the collection,
as are several acorn cups. The abundance of this oak in the flora is understandable, for
the modern species most allied to it, *Q. chrysolepis* Liebmann, prefers moist sites and
is a frequent inhabitant of streambanks. It ranges from the broadleaved sclerophyll
forest up into mixed conifer forest in open, well-drained sites.
   Regarding the abundance of entire-margined leaves in the flora, I have noted that the
frequency of serrate leaves in modern stands decreases with elevation and with
increasingly cooler conditions, a relation consistent with the composition of the fossil
flora.
   Occurrence: Buffalo Canyon, hypotypes 9519-9521, 9523, 5924, 9546, 9547;
homeotypes 9522, 9525-9539.

Quercus wislizenoides Axelrod
(Plate 12, figs. 11-12)

Quercus wislizenoides Axelrod, Carnegie Inst. Wash. Pub. 553, p. 136, pl. 29, figs. 4-9,
      1944.
   Axelrod, Univ. Calif. Pub. Geol. Sci. 33, p. 291, pl. 14, figs. 1, 2; pl. 20, figs. 4,
      5, 8; pl. 27, figs. 5-8, 1956.
   Axelrod, Univ. Calif. Pub. Geol. Sci. 121, p. 111, pl. 11, figs. 1-7, 1980a.

A dozen leaves in the flora are identified as this species on the basis of the
irregularly branching secondaries and the coarse network of 4th- and 5th-order veinlets.
These features contrast with those of *Q. hannibalii* (cf. *chrysolepis*), in which the
secondaries scarcely branch and are not irregular, and the finer meshwork of veins is
much denser.
   The fossil species seems most nearly allied to the modern *Q. wislizenii* DeCandolle,
a common member of the oak woodland and sclerophyll forest of California. *Q.
wislizenoides* probably was confined to the warmest, driest sites provided by volcanic
rocks bordering the lake.
   Occurrence: Buffalo Canyon, hypotypes 9546-9549, homeotypes 9550-9552.

Family ULMACEAE
Ulmus

Revision of the Neogene elm species in western North America by Tanai and Wolfe (1977) raises a number of problems that result in further confusion with respect to species identification. The characters that they selected for separation of species are not consistent, nor were they applied consistently. Some of these are noted in the following discussion of *Ulmus speciosa*.

Ulmus speciosa Newberry
(Plate 12, figs. 8-10; pl. 13, fig. 10)

Ulmus speciosa Newberry, U.S. Geol. Surv. Monogr. 35, p. 80, pl. 45, figs. 3, 4, 1898.

    Tanai and Wolfe, U.S. Geol. Surv. Prof. Paper 1026, p. 8, pl. 3, figs. C, F, 1977 (see synonymy).

Ulmus owyheensis Smith, Michigan Acad. Sci. Papers, Arts and Letters 24, p. 113, pl. 6, fig. 4, 1939.

    Tanai and Wolfe, U.S. Geol. Surv. Prof. Paper 1026, p. 6, pl. 2, figs. B, D, E, F; pl. 3A, 1977.

Several leaves in the Buffalo Canyon flora are referred to this species on the basis of their similarity to the type specimens. Although the leaves have a prominent tertiary vein that directly enters the sinus, the teeth are both acutely trigonal and bluntly deltoid, and hence cannot be identified in terms of the criteria presented by Tanai and Wolfe (1977); similar variation is present in other specimens that they include in the species. The leaves of *U. owyheensis* Smith cannot be separated from the illustrated types of *U. speciosa*; especially compare the excellent specimen figured by Tanai and Wolfe (pl. 3, fig. A) with the types of *U. speciosa*. Specimens of *U. owyheensis* are therefore placed in synonymy.

Furthermore, in some leaves the base is slightly asymmetrical, in others strongly asymmetrical, defying their key to identification. In view of these features, I prefer to recognize one species, not two or three—or a new one—as would be demanded if we adhered strictly to the criteria of Tanai and Wolfe. In this regard, I have just examined (Oct. 13, 1986) the leaves that are commencing to fall from *U. glabra* and *U. procera* along the central campus road at Davis. They clearly show that any species has leaves that vary considerably in shape and marginal features. The description of such variation in fossil suites only proliferates "species" and does not necessarily enhabce our understanding of fossil elms, alders, maples, or other taxa.

Apart from the leaves, there are 36 small samaras (12-15 mm long) in the collection. All of them have similar features and are of essentially the same size, implying that only one species of elm is probably represented in the flora. They are slightly smaller than those produced by *U. americana* Linnaeus of the eastern United States. However, the leaves of *U. speciosa* differ from those of *U. americana* in that the teeth of the fossil leaves are sharply trigonal on one side of the blade but bluntly deltoid on the other. By contrast, *U. americana* has sharply trigonal teeth on both sides of the leaves. It thus appears that *U. speciosa* is extinct.

Occurrence: Buffalo Canyon, hypotypes 9553-9555, homeotypes 9556-9564.

Zelkova brownii Tanai and Wolfe
(Plate 13, figs. 5-9)

Zelkova brownii Tanai and Wolfe, U.S. Geol. Surv. Prof. Paper 1026, p. 8, pl. 4, figs. A, C-G, 1977 (see synonymy and discussion).

Zelkova leaves are among the most common leaf fossils at the Buffalo Canyon site. They display great variation in size and shape, ranging from ovate to broadly elliptic to narrowly so, and from rounded to asymmetrical at the base. The larger leaves are 10-14 cm long, but most are 5-7 cm long. The margins are characterized by large, simply dentate, often apiculate teeth, and some teeth may have a subsidiary serration. The finer nervation is well preserved and shows that the tertiaries are weak and irregular, not strong and subpercurrent as in *Ulmus*.

This species has usually been compared with *Z. serrata* (Thunberg) Makino and *Z. sinica* Schneider of Japan and China, though Tanai and Wolfe note that the teeth of *C. carpinifolia* of the Caucasus Mountains are more like the fossils. However, important differences are that both *serrata* and *brownii* have 8-14 pairs of secondaries, whereas *carpinifolia* has 6-8 pairs; the secondaries of the Oriental species are firm and straight, unlike the more irregular secondaries of *carpinifolia*; leaves of *carpinifolia* are usually smaller, and the teeth of *brownii* and the Oriental species are regularly apiculate. The fact that some of the leaves of *brownii* in the Pacific Northwest floras have teeth that are not markedly apiculate implies that an ancient character was still present in their makeup. Certainly, among living species, *Z. brownii* is more nearly allied to the Oriental *serrata* and *sinica* than to *carpinifolia*.

Occurrence: Buffalo Canyon, hypotypes 9565-9568, 9572; homeotypes 6870, 9569-9571, 9573-9581.

Family BERBERIDACEAE

Mahonia macginitiei Axelrod
(Plate 14, figs. 3-4, 7, 9)

Mahonia macginitiei Axelrod, Univ. Calif. Pub. Geol. Sci. 129, p. 150, pl. 11, figs. 2, 4, 9; pl. 27, fig. 7, 1985 (see synonymy and discussion).

Leaflets of this species are common in the Buffalo Canyon flora and exhibit much of the variation of the species. They are ovate to elliptic with a subequal base. The margins are shallowly dentate with sharp spines, the number varying with position on the rachis.

Similar leaflets are produced by *M. aquifolium* (Pursh) Nuttall, distributed in the coastal area from central California northward into British Columbia, with a salient into the Idaho Panhandle where other coastal species also reappear.

Occurrence: Buffalo Canyon, hypotypes 9582-9585, homeotype 9586.

Mahonia reticulata (MacGinitie) Brown
(Plate 14, figs. 5, 6, 8)

Clematis reticulata MacGinitie, Carnegie Inst. Wash. Pub. 416, p. 45, 1933.
Mahonia reticulata (MacGinitie) Brown, U.S. Geol. Surv. Prof. Paper 186J, p. 175 (name only), 1937.
  Axelrod, Univ. Calif. Pub. Geol. Sci. 129, p. 152, pl. 11, figs. 5, 10; pl. 26, figs. 1-5; pl. 27, figs. 1-2, 1985 (see synonymy and discussion).

Leaflets of this species are well represented in the Buffalo Canyon flora. They are elliptic to ovate-elliptic, characterized by broadly spaced marginal spine. The leaflets resemble those of *M. insularis* Munz of the Channel Islands off Santa Barbara, where it forms a climbing shrub-vine. The leaflets also resemble those of some Mexican species, notably *M. andrieuxii* (H. and A.) Fedde, *M. ochochoco* (Schindl.) Fedde, and *M. longipes* (Standley) Standley. This relationship is consistent with the composition of the present woody flora of coastal southern California. A number of its taxa have their nearest allies in Mexico, and during the Neogene numerous species presently in Mexico had allies that were then in the Great Basin and southern California.
  Occurrence: Buffalo Canyon, hypotypes 9587, 9588, 9590; homeotypes 9589, 9591-9593.

Family NYMPHAEACEAE

Nymphaeites nevadensis (Knowlton) Brown
(Plate 17, figs. 5, 6)

Nymphaeites nevadensis (Knowlton) Brown, Wash. Acad. Sci. Jour. 27, p. 509, fig. 10, 1937 (see synonymy).
  Chaney and Axelrod, Carnegie Inst. Wash. Pub. 617, p. 175, pl. 32, figs. 8, 9; pl. 33, figs. 9, 10, 1959.
  Axelrod, Univ. Calif. Pub. Geol. Sci. 129, p. 156, pl. 11, figs. 6-7; pl. 22, fig. 9; pl. 28, figs. 1, 3-4; pl. 29, fig. 9, 1985.

The record of this species is based on detached root scars of rhizomes and fragmentary leaf impressions, some of which are 15-18 cm long. The present material compares well with the leaves of *Nuphar polysepalum* Engelmann, widely distributed in ponds and lakes from central California north into Alaska.
  More complete material is needed before it would be possible to separate the two or three species from those that are now grouped under *Nymphaeites*.
  Occurrence: Buffalo Canyon, hypotypes 9594-9596, homeotypes 9597-9600.

Family HYDRANGEACEAE

Hydrangea bendirei (Ward) Knowlton
(Plate 15, figs. 7-9)

Hydrangea bendirei (Ward) Knowlton, U.S. Geol. Surv. Bull. 204, p. 60, pl. 9, figs.
    6, 7, 1902.
        Chaney and Axelrod, Carnegie Inst. Wash. Pub. 617, p. 180, pl. 36, figs. 10, 11
    (not fig. 9, which is *H. ovatifolia*), 1959.

Three slender, elliptic leaves in the flora are similar to the large suite of *H. bendirei*
leaves in the Mascall flora. They resemble leaves of several east Asian species, notably
H. aspera Don and *H. chinensis* Maximowicz. *H. bendirei* has been recorded at a
number of sites in Oregon and Idaho, as tabulated by Chaney and Axelrod (1959, p.
180).
Occurrence: Buffalo Canyon, hypotypes 9601-9603.

Family GROSSULARIACEAE

Ribes barrowsae Axelrod, n. sp.
(Plate 15, figs. 2-4)

Description: Leaf palmately 3-lobed; blade 2.5 cm long, 2.4-2.6 cm wide; petiole
2.5 cm long, 1 mm thick; primary veins of each lobe thinning distally, alternate
secondaries diverging nearly parallel to midrib, entering marginal teeth, or supplying
them with tertiaries; tertiaries forming a series of broad loops within the margin of each
lobe; margin with a few remote, rounded to blunt teeth; texture medium.
Discussion: These leaves resemble those of *R. aureum* Pursh, a common streambank
shrub in the Great Basin and border areas, reaching from the sagebrush belt well up into
the lower margin of subalpine forest and ranging east to the Rocky Mountains.
    R. auratum Becker (1961, pl. 20, fig. 13) was compared with *R. aureum*. However,
the tips of the leaf lobes are acute rather than rounded, the base is cuneate, not truncate,
and the petiole is short. The leaf of *R. auratum* Becker is inseparable from *Holodiscus
idahoensis* Chaney and Axelrod, the leaves of which are especially abundant in the
Creede flora of Colorado (Axelrod, 1987).
    This species is named for Katherine J. Barrows, who mapped the geology of the
Buffalo Basin and the bordering Eastgate Hills and Desatoya Mountains (Barrows, 1971).
Occurrence: Buffalo Canyon, holotype 9604, paratypes 9605-9607.

Ribes bonhamii Axelrod, n. sp.
(Plate 15, fig. 1)

Description: Leaf palmately 3-lobed, the lobes elliptic-rhombic, acute above; leaf
blade 3.0 cm long, 3.2 cm broad; petiole relatively thin, 2.0 cm long; midrib of the
lateral lobes slightly arched; secondaries departing at low angles and reaching up to

relatively large, bluntly rounded teeth which may have subsidiary blunt teeth supplied by tertiaries; thin tertiaries form loops within the margin of the leaf lobes; texture medium.

Discussion: No previously described fossil *Ribes* species is similar to *R. bonhamii*, which shows relationship to leaves of *R. americanum* Miller of the eastern United states and also to *R. meyeri* Maximowicz of the mountains of China. Although there are numerous species of *Ribes*, in the western United States, none seem to have leaves as similar to the fossil as the eastern species. Nearly all of the 31 species in California have principal leaf lobes that are broadly rounded, not acute as in the fossil.

It is a pleasure to name this species for Harold F. Bonham, geologist of the Nevada Bureau of Mines, who has assisted in collecting rock samples for radiometric dating and who has generously devoted field time to discussing various petrologic and structural problems in this area and elsewhere in western Nevada and adjacent California.

Occurrence: Buffalo Canyon, holotype 9608, paratype 9609.

Ribes stanfordianum Dorf
(Plate 15, fig. 10)

Ribes stanfordianum Dorf, Carnegie Inst. Wash. Pub. 412, p. 97, pl. 10, fig. 6, 1930.
   Axelrod, Univ. Calif. Pub. Geol. Sci. 39, p. 234, pl. 48, figs. 5-6, 1962.
   Axelrod, Univ. Calif. Pub. Geol. Sci. 129, p. 159, pl. 29, figs. 1, 3, 1985.

A single specimen in the flora is broadly 3-lobed and has all the characters typical of currant. In view of the limited material, it is not easy to designate any one living species as most similar to the fossil. Nonetheless, leaves of *R. nevadense* Kellogg and *R. viscosissimum* Pursh closely resemble the fossil.

Occurrence: Buffalo Canyon, hypotype 9610.

Ribes webbii Wolfe
(Plate 15, figs. 5, 6)

Ribes webbii Wolfe, U.S. Geol. Surv. Prof. Paper 454-N, p. N23, pl. 9, figs. 13, 14, 17, 18, 1964a.

Several leaves in the flora are similar to those recorded from the Stewart Spring flora. They are palmate, broadly orbicular, with a rounded apex and truncate to cordate base. The marginal lobes are crenulate, with small rounded teeth.

These leaves resemble those of *R. cereum* Douglas, a widespread shrub in the western United States, ranging from the piñon-juniper belt well up to the subalpine forest zone. *R. cereum* prefers drier sites, and *R. webbii* may be presumed to have occupied exposed, well-drained volcanic slopes bordering the lake.

Occurrence: Buffalo Canyon, hypotype 9611, 9613; homeotype 9612.

Family ROSACEAE

Amelanchier desatoyana Axelrod, n. sp.
(Plate 16, figs. 1-7)

Amelanchier grayi Chaney. Axelrod, Univ. Calif. Pub. Geol. Sci. 129, p. 159, pl. 28,
fig. 2; pl. 30, figs. 2, 3, 1985.

Description: Leaves broadly ovate to orbicular or (rarely) elliptic, 1.8-4.0 cm long,
typically 3.0-3.5 cm wide; slender petiole, 1.0-1.5 cm long; apex rounded to bluntly
obtuse, base rounded to obtuse; about 7 alternate to subopposite secondaries, diverging
at moderate angles, looping upward just inside margin in lower half of blade, then either
entering teeth directly or dividing into tertiaries; secondaries recurving toward apex in
distal part of blade; tertiary venation in lower third strongly cross-percurrent, becoming
more irregularly polygonal above, the polygons of quadrilles enclosing an open-ended,
very dense net of veinlets; very acute teeth in upper half to third of blade; texture
medium.

Discussion: The excellently preserved leaves in the Buffalo Canyon flora show more
clearly that they, as well as those in the Eastgate flora, represent a distinct new species
of *Amelanchier*.

These leaves differ from those of *A. alvordensis* Axelrod (1944b) which, based on
larger collections, are commonly obovate and do not have such a strong development of
cross-percurrent tertiaries as the Buffalo Canyon specimens. In *A. grayii* Chaney from
Gray Ranch, Oregon (Chaney, 1927), the percurrency of the tertiaries is not as
pronounced; further study may show that it and *A. alvordensis* Axelrod (1944b) are
similar.   *A. scudderi* Cockerell (MacGinitie, 1953) from Florissant has fewer
secondaries, and they are not as looping; the tertiaries are more irregular and trend
normal to the secondaries, and the teeth are more rounded.

These fossil leaves are similar in most respects to those of the modern *A. florida*
Lindley of the northern coast of California, ranging north to Alaska. Specimens from
the Rocky Mountains, usually identified as *A. alnifolia* Nuttall, also resemble the fossils.
It is noted, however, that the fossil species appears to have a much denser groundwork
of finer nervilles.

Occurrence: Buffalo Canyon, holotype 9614, paratypes 9615-9620.

Cercocarpus ovatifolius Axelrod
(Plate 16, fig. 8)

Cercocarpus ovatifolius Axelrod, Univ. Calif. Pub. Geol. 129, p. 163, pl. 32, figs. 5-
13, 1985.

A single leaf of this species was collected. Although it is not so ovate as the type,
it falls readily within the variation displayed by the large suite of specimens in the
Eastgate flora. This species is most nearly allied to *C. blancheae* Schneider of the
Channel Islands that also occurs at a few relict sites in the Santa Monica Mountains, of
southern California. The relationship of *C. ovatifolius* to other fossil species is discussed
elsewhere (Axelrod, 1985) and need not be reviewed here.

Occurrence: Buffalo Canyon, hypotype 9621.

Chamaebatia nevadensis Axelrod, n. sp.
(Plate 17, fig. 4)

Description: Lateral leaf blade elliptic, 1.7 cm long, 0.5 cm broad in upper third, narrowing basally to 1 mm broad segments; petiole 3 mm long, thick; blade with 10 alternate secondary branchlets varying from 1 mm to 2.5 mm long, broadly rectangular in outline to somewhat obovate, broader segments with 4 to 5 divisions or "lobes", which are subopposite, entire, and broadly rounded distally; the terminal lobe frequently club-shaped, each lobe adnate on the ternary laminar division; venation not visible, possibly obscured by glutinous secretions, as in the modern shrub; texture firm and apparently thick.

Discussion: This specimen is similar to the foliage of *C. foliolosa* Bentham. This is a low shrub, from 15-25 cm tall. The intertwining strong branches of the individual shrubs are mixed with those of adjacent plants to form a nearly continuous mat on the forest floor. The species occurs in open or semi-open areas in the yellow pine forest on the western slope of the Sierra Nevada.

Occurrence: Buffalo Canyon, holotype 9710.

Crataegus middlegatei Axelrod
(Plate 16, figs. 9, 10)

Crataegus middlegatei Axelrod, Univ. Calif. Pub. Geol. Sci. 33, p. 300, pl. 29, figs. 3, 4, 1956.

Three fragmentary leaves in the Buffalo Canyon flora represent this species. They are similar to those produced by *C. chrysocarpa* Ashe and *C. columbiana* Howell. Leaves of the present Rocky Mountain species, *C. coloradensis* A. Nelson and *C. erythropoda* Ashe are also generally similar to the fossil leaves. All these species are large shrubs and in general prefer moist sites. They are commonly found in riparian situations and hence are in a favorable position to contribute to an accumulating fossil record.

Occurrence: Buffalo Canyon, hypotype 9622, 9623; homeotype 9624.

Heteromeles desatoyana Axelrod, n. sp.
(Plate 16, fig. 11)

Photinia sonomensis Axelrod, Carnegie Inst. Wash. Pub. 553, p. 258, pl. 45, fig. 1, 1944b.
Heteromeles sonomensis (Axelrod) Axelrod, Univ. Calif. Pub. Geol. Sci. 129, p. 166, pl. 12, figs. 7, 8, 10; pl. 30, figs. 1, 4, 6, 1985.

Description: Leaves long-elliptic to slightly obovate; 7.0-8.0 cm long and 3.0-3.5 cm wide; tip acute to rounded, base acute; midrib stout, tapering above; petiole thick and heavy, about 1 cm long; numerous secondaries diverging at about 45°, looping well within margin to join secondaries above, sending tertiaries to marginal teeth; tertiary mesh irregularly polygonal; the finer quaternary mesh somewhat elongated roughly subparallel to the secondaries, enclosing a finer intercostal mesh with open cross-ties; margin coarsely serrate, the teeth widely spaced, often recurved, apparently gland-tipped.

Discussion: Specimens in the Buffalo Canyon flora, as well as in the Alvord Creek (Axelrod, 1944b) and the Middlegate and Eastgate (Axelrod, 1985) floras, are consistently much larger than those referred previously to *H. sonomensis* Axelrod. They display all the typical features of large-leaved *H. arbutifolia* Roemer, an evergreen shrub or small tree in California that ranges up to the forest margin. The specimens differ from leaves of *Photinia*, which are more finely serrate and have a denser, more complex and finer nervation.

Occurrence: Buffalo Canyon, holotype 9625, paratypes 9626-9628; Alvord Creek, hypotype 2118, homeotype 2119; Middlegate, hypotypes 6622-6624; Eastgate, hypotypes 7063-7066.

Lyonothamnus parvifolius (Axelrod) Wolfe
(Plate 18, figs. 1, 2)

Lyonothamnus parvifolius (Axelrod) Wolfe, U.S. Geol. Surv. Prof. Paper 454-N, p. N26, 1964a (discussion only).
　　Axelrod, Univ. Calif. Pub. Geol. Sci. 129, p. 167, pl. 12, figs. 1, 4, 9; pl. 27, fig. 8; pl. 29, fig. 8, 1985.

Subsequent collections of nearly complete leaves show that these fossils represent *Lyonothamnus*, the Catalina ironwood, not *Comptonia* (Axelrod, 1956). It is noteworthy that specimens from the central Great Basin floras, such as Aldrich Station, Eastgate, Middlegate, Purple Mountain, and Buffalo Canyon, have much smaller leaves, and the leaflet lobations are also much smaller, than those of the Stewart Spring fossils illustrated by Wolfe (1964a). The consistently larger size of the Stewart Spring specimens indicates that they represent a different species, *L. cedrusensis* Axelrod (the type of which is illustrated in Wolfe, 1964, pl. 11, fig. 5; the other specimens figured are paratypes). This large-leafed species may well reflect the somewhat warmer climate at the south in the Cedar Valley area. *L. cedrusensis* also has fewer and slenderer leaf lobes than the modern *L. asplenifolius* of the Channel Islands. The development of broader, and presumably thicker, leaf lobes of the present insular endemic may reflect adaptation to a humid, equable coastal climate following the later Miocene.

Occurrence: Buffalo Canyon, hypotypes 9629, 9630; homeotypes 9631-9634.

Prunus chaneyii Condit
(Plate 17, figs. 1, 3)

Prunus chaneyii Condit, Carnegie Inst. Wash. Pub. 476, pl. 5, figs. 4, 5, 1938.
　　Chaney and Axelrod, Carnegie Inst. Wash. Pub. 617, p. 185, pl. 36, fig. 4, 1959.
　　Axelrod, Univ. Calif. Pub. Geol. Sci. 51, p. 124, pl. 14, figs. 5-8, 1964.
　　Axelrod, Univ. Calif. Pub. Geol. Sci. 129, p. 168, pl. 30, fig. 9, 1985.

Six specimens represent this species, whose leaves resemble those of the *P. demissa* (Nuttall) Walpers-*melanocarpa-virginiana* Linnaeus phylad. The fossils are broadly elliptic with acuminate tips. The margins are finely serrate, and the camptodromous secondaries loop well within the margin and fork to supply the marginal area with tertiaries that divide once or twice to diverge into the fine serrate teeth.

The modern species allied to the fossil are chiefly forest inhabitants which descend along streams to lower levels. *P. demissa* occurs most frequently at seepages or springs throughout the Great Basin region, from the sage belt up through the conifer woodland into the forest zone.

Occurrence: Buffalo Canyon, hypotypes 9635, 9636; homeotypes 9637, 9638.

Prunus moragensis Axelrod
(Plate 17, fig. 2)

Prunus moragensis Axelrod, Carnegie Inst. Wash. Pub. 553, p. 140, pl. 30, figs. 6, 8, 10, 1944.
Axelrod, Univ. Calif. Pub. Geol. Sci. 33, p. 301, pl. 28, fig. 15, 1956.
Axelrod, Univ. Calif. Pub. Geol. Sci. 34, p. 132, pl. 23, figs. 11-13, 1958.
Axelrod, Univ. Calif. Pub. Geol. Sci. 39, p. 235, pl. 49, fig. 5, 1962.

A single, oblong-elliptic leaf in the flora has numerous alternate secondaries that diverge at moderate angles, loop upward close to the margin, and then divide into tertiaries that supply numerous crenately-serrate teeth. Similar leaves are produced by bitter cherry, *P. emarginata* (Douglas) Walpers but differ in that the secondaries in the modern species loop well within the blade, farther from the margin than in the fossil. This may reflect its ecology, with the moister climate of the Miocene favoring a more marginal position for the secondaries.

P. emarginata is a shrub or small tree distributed from sea level to high montane sites, ranging from California north into British Columbia and east into Idaho.

Occurrence: Buffalo Canyon, hypotype 9639.

Prunus treasheri Chaney
(Plate 17, figs. 7-9)

Prunus treasheri Chaney, Carnegie Inst. Wash. Pub. 553, p. 347, pl. 63, fig. 3; pl. 64, fig. 1, 1944.

Well-preserved leaves of this species in the Buffalo Canyon flora are lanceolate, with an acuminate tip and rounded base. Numerous secondaries diverge at moderate angles, looping well within the margin to supply the sharply serrate teeth with tertiaries. Within the blade, the tertiaries form a quadrangular net that encloses a mesh of finer veinlets, some of which have open cross-ties. Texture is medium.

These leaf impressions are similar to leaves of the native peach of northern China, *P. davidiana* (Carrière) Franchet, the chief difference being that *P. davidiana* leaves have a more acuminate apex.

Occurrence: Buffalo Canyon, hypotypes 9640, 9642, 9643; homeotype 9641, 9644.

Rosa harneyana Chaney and Axelrod
(Plate 18, figs. 3-5)

Rosa harneyana Chaney and Axelrod, Carnegie Inst. Wash. Pub. 617, p. 186, pl. 37,
    fig. 4, 1959 (see synonymy).
Prunus moragensis Axelrod. Chaney and Axelrod, ibid, p. 185, pl. 36, fig. 5, 1959.

Several well-preserved asymmetrical leaflets elliptic in outline, acute above and
cuneate to blunt below are referred to this species. The finer details of nervation are
well preserved and clearly those of a rose. The species differs from *R. hilliae*
Lesquereux in the Florissant flora, which has widely spaced, simply dentate or rarely
doubly dentate teeth, whereas Buffalo Canyon specimens have sharply serrate or
occasionally biserrate teeth.

The Trout Creek specimen of *Prunus* listed above was incorrectly illustrated. Since
it has looping secondaries, it cannot be a *Prunus* allied to *emarginata*. In addition, the
cherry has crenate teeth, not sharply serrate ones like the Trout Creek specimen, whose
asymmetry also suggests that it is a *Rosa* leaflet.

Leaves figured as *R. hilliae* Lesquereux from the Ruby flora (Becker, 1961, pl. 23)
are generally smaller than the present fossils, and the serration more nearly resembles
that of the Florissant material.

The present leaflets compare favorably with those of *Rosa nutkana* Presley of the
Pacific Northwest, and also *R. setigera* Michaux of the eastern United States.

Occurrence: Buffalo Canyon, hypotypes 9645, 9647, 9648; homeotypes 9646, 9649.

Sorbus cassiana Axelrod
(Plate 18, figs. 6-8)

Sorbus cassiana Axelrod, Univ. Calif. Pub. Geol. Sci. 51, p. 125, pl. 14, fig. 3 only
    (not fig. 4, which is *Alnus*), 1964.
Sorbus idahoensis Axelrod, Univ. Calif. Pub. Geol. Sci. 129, p. 169, pl. 29, figs. 2, 6-
    7, 10, 1985.
Rhus alvordensis Axelrod, Univ. Calif. Pub. Geol. Sci. 33, p. 305, pl. 30, fig. 8, 1956.
    Axelrod, Univ. Calif. Pub. Geol. Sci. 51, p. 125, pl. 14, fig. 11, 1964.

This species is distinguished by slender leaflets arranged oppositely on pinnate
leaves, and the margins have rather widely spaced, single-serrate teeth. It thus differs
from *S. alvordensis* Axelrod, which is biserrate (Axelrod, 1944b, pl. 44, figs. 6, 7).
The excellently preserved specimens from Buffalo Canyon are similar to the leaflets of
Sorbus pohuashanensis (Hance) Hedley of northern China and also to those of *S.
aucuparia* Linnaeus of Eurasia.

S. idahoensis from the Eastgate flora differs in no fundamental way from *S. cassiana*
and is therefore reduced to synonymy. *S. cassiana* (Axelrod, 1964, pl. 14, fig. 3) has
priority.

Occurrence: Buffalo Canyon, hypotypes 9650-9652, homeotypes 9658, 9659.

Family FABACEAE

Amorpha stenophylla Axelrod, n. sp.
(Plate 19, figs. 5, 6)

Description: Leaflets narrowly lanceolate, tip apiculate, base slightly asymmetrical, rounded; 3.2 cm long and 0.7 cm broad in lower third; petiolule missing; 8-10 alternate secondaries, camptodrome; strong intersecondaries; tertiaries form a coarse, irregularly elongated polygonal network enclosing a series of irregular quaternary polygons; margin entire; texture firm.

Discussion: Two leaflets appear to represent a fossil species allied to *A. angustifolia* Pursh of Texas, Kansas, Arkansas, and border areas.

The fossil differs from previously described fossil species in its slender outline. *A. oblongifolia* Axelrod has an oblong outline and is allied to *A. californica* Nuttall. *A. oklahomensis* (Brown) Axelrod has larger elliptic leaves and is more nearly allied to *A. fruticosa* Linnaeus, widely distributed from California eastward.

Occurrence: Buffalo Canyon, holotype 9660, 9660a, paratype 9661.

Robinia bisonensis Axelrod, n. sp.
(Plate 19, fig. 12)

Description: Leaflet elliptic, 5 cm long and 2.6 cm broad; apex broadly rounded, with a small mucro; base broadly acute; petiolule missing; margin entire; midrib firm, straight, and thinning above; 7 alternate secondaries diverging at medium angles, camptodrome; tertiaries form a coarse polygonal network enclosing coarse 4th-order veinlets that form polygons of irregular shape and enclose a 5th-order mesh of very dense veinlets; texture medium thin.

Discussion: A single large, well-preserved leaflet appears to represent *Robinia* and is very similar to those produced by *R. pseudoacacia* Linnaeus. This medium-sized tree inhabits the moister parts of the Appalachian and Ozark Mountains.

This fossil species may be similar to that figured as *Phyllites sophoroides* (Knowlton, 1926, pl. 26, fig. 8). However, until all the leguminous leaflets in the Latah flora are revised, it seems best to keep the Buffalo Canyon locust separate.

Occurrence: Buffalo Canyon, holotype 9662.

Family ACERACEAE

Acer medianum Knowlton
(Plate 19, fig. 11)

Acer medianum Knowlton, U.S. Geol. Surv. Bull. 102, p. 76, pl. 14, figs. 4, 5, 1902.
  Wolfe and Tanai, Jour. Fac. Sci. Hokkaido University 22, p. 111, pl. 38, figs. 1-6;
  pl. 39, figs. 1-9; pl. 40, figs. 1-5, 7, 8, 11; text-fig. 14N, 1987 (see synonymy and discussion).

A single small trilobed leaf represents this species which, according to Wolfe and Tanai, is a member of the extinct Section *Columbiana*. As their synonymy indicates, there is considerable variation in the shape of the leaves and degree of incision. In general habit, the species may have been similar to *A. glabrum*, though only distantly related to it.

Occurrence: Buffalo Canyon, hypotype 9664, 9664a.

Acer negundoides MacGinitie
(Plate 19, figs. 4, 9, 10, 13)

Acer negundoides MacGinitie, Carnegie Inst. Wash. Pub. 416, p. 62, pl. 11, figs. 2-3, 1933.
    Axelrod, Univ. Calif, Pub. Geol. Sci. 129, p. 172, pl. 14, figs. 7-9, 1985 (see synonymy and discussion).

The typical distinctive samaras of this species are abundant in the collection, and several leaflets are also present. They show relationship to box elder, *A. negundo* Linnaeus, a widely distributed tree in riparian sites across much of central North America. Relationship is also apparent with *A. henryii* Pax of central China.

Occurrence: Buffalo Canyon, hypotypes 9665-9670, 9894; homeotypes 9671-9676.

Acer oregonianum Knowlton
(Plate 19, figs. 1-3)

Acer oregonianum Knowlton, U.S. Geol. Surv. Bull. 204, p. 75, pl. 13, figs. 5, 7, 8, 1902.
    Chaney and Axelrod, Carnegie Inst. Wash. Pub. 617, p. 195, pl. 41, figs. 11-14, 1959 (see synonymy).
    Axelrod, Univ. Calif, Pub. Geol. Sci. 129, p. 175, pl. 13, figs. 8-9; pl. 15, figs. 4-5; pl. 34, figs. 1-2, 1985.

The samaras of this species are well represented in the Buffalo Canyon flora, though remains of its leaves have not been recovered. The samaras are very distinctive, being quite large and with a typical constriction or bay in the wing under the seed.

Among living species, the fossils are quite similar to samaras of the bigleaf maple, *A. macrophyllum* Pursh. This large tree inhabits moist valleys and stream banks, distributed from southern California northward into coastal Canada and west of the Sierra-Cascade axis.

In Section *Macrophylla*, Wolfe and Tanai (1987) recognized several new entities among fossils previously referred to *A. oregonianum* Knowlton, which is allied to *A. macrophyllum* Pursh. That these "new" species probably represent no more than ecologic variants is suggested by the climatically correlated regional differences in samara size (the commonest fossil structure in *Acer*) of *A. macrophyllum*. Specimens from interior southern California and the interior lowland valleys of central California generally range from 3.0-3.5 cm long; those from the mixed conifer forest of the central Sierra Nevada and north Coast Ranges vary from 4.0-4.5 cm, while most specimens from the humid, more equable coastal strip that extends north to British Columbia range from 4.5-6.5 cm. These figures differ from those presented by Wolfe and Tanai (p. 141, Table 8) because they grouped all 100 measurements of *macrophyllum* samara length into a mean of 4.7 cm and a range of 3.4-6.3 cm. Actually, many specimens from interior southern California are only 3.0 cm long.

Analysis of several fossil collections suggests that samara size also varied with climate. The small samaras of the Salmon, Alvord Creek, and Trapper Creek floras reflect the cool temperate climate of those conifer-rich forests. The Thurston Ranch and Pyramid floras, with samaras averaging 6.7-7.6 cm long, respectively, reflect the environment of deciduous hardwood forests with ample summer rain and mild temperature. Samaras from the Fingerrock flora (average 6.4 cm) are somewhat smaller, probably owing to the warmer climate and higher evaporation there. Wolfe and Tanai recognized var. *busamarum* on the basis of its large samaras. Var. *busamarum* has an average length of 7.5 cm, whereas var. *fingerrockense* has samaras averaging 6.3 cm. How are we to treat the samaras from Thurston Ranch and Pyramid floras that are intermediate between these varieties? Are they to be designated a "new" intermediate variety of *busamarum*? I do not believe that this is sound systematics. The differences reflect local environments and hence variation in the diverse minor characters that were selected for specific status by Wolfe and Tanai. Furthermore, the problem is compounded in cases in which the record is represented by only 1 or 2 samaras. Do these represent the average size of the taxon at that site? Probably not. It is also apparent that if climate for a particular year or set of years is drier, colder, or warmer than normal, it may also affect size, much as it does with cones of pine, fir, spruce, and no doubt others.

In this regard, the samaras in the Buffalo Canyon flora occur in two kinds of matrix. Those from a dense tuffaceous mudstone are of medium size (average 4.5 cm), whereas two from white diatomaceous matrix are much larger, measuring 6.5 and 7.5 cm, respectively (specimens 9893 and 9901). Are these to be regarded as different varieties? Since they occur stratigraphically within 10-15 cm of one another, it seems more likely that they may reflect local climatic change, from cooler to slightly warmer conditions. Rather than recognize these as variants, I believe that it is sounder systematics to refer all the material to a generally recognizable species, *Acer oregonianum*, which all agree is allied to *Acer macrophyllum*.

Occurrence: Buffalo Canyon, hypotypes 9678, 9679, 9693; homeotypes 9680, 9681, 9901.

Acer trainii Wolfe and Tanai
(Plate 19, fig. 8)

Acer trainii Wolfe and Tanai, Jour. Fac. Sci. Hokkaido University 22, p. 81, pl. 37, figs. 20, 21, 1987.

A single samara in the flora is referred to this species on the basis of the prominent longitudinal crease along the middle of the seed and the smaller ones bordering it. The samara is 2.5 cm long, and the nutlet is 9 cm long and 5 cm wide. The allied *A. glabrum* Torrey regularly has similar samaras. This large shrub is common in the mesic western forests, where it is represented by several varieties.

Occurrence: Buffalo Canyon, hypotype 9677.

Acer tyrellii Smiley
(Plate 19, fig. 7)

Acer tyrellii Smiley, Univ. Calif. Pub. Geol. Sci. 35, p. 227, pl. 13, figs. 3, 5, 1963.
    Wolfe and Tanai, Jour. Fac. Sci. Hokkaido University 22, p. 180, pl. 55, figs. 7-
    11; pl. 56, figs. 1-4; text-fig. 18F, 1987 (see synonymy).

The terminal portion of a samara with seed complete and most of the wing missing
is similar to those produced by the living *A. grandidentatum* and *A. brachypterum* of the
southern Rocky Mountains.

In discussing species of this alliance, Wolfe and Tanai included two very different
samaras in *A. schornii*. On Plate 55, figs. 3 and 5 are similar, but fig. 4 has a much
steeper angle (45°) of attachment and seems more properly referred to *A. tyrellii*. The
present specimen from Buffalo Canyon is similar to it, although somewhat smaller in
size. The specimen is readily matched by samaras of the modern *A. grandidentatum* as
well as *brachypterum*. Variation in samara size appears to be no more than a reflection
of the local habitat—cool and moist, or warm and subhumid.

Occurrence: Buffalo Canyon, hypotype 9682.

Family CAPRIFOLIACEAE

Symphoricarpos wassukana Axelrod
(Plate 21, figs. 4, 5)

Symphoricarpos wassukana Axelrod, Univ. Calif. Pub. Geol. Sci. 33, p. 312, pl. 9, fig.
    2 only, 1956.

A single small ovate leaf in the Buffalo Canyon flora has thin, camptodromous
secondaries that diverge at moderate angles. The thin tertiary venation forms an
irregular polygonal pattern that encloses a coarse open polygonal network of finer veins.
This specimen is similar to leaves of *S. oreophilus* Gray and *S. vaccinoides* Rydberg
(=*S. rotundifolius* Gray) of the montane forests of the western United States.

Occurrence: Buffalo Canyon, hypotypes 9708, 9708a, 9709.

Family MELIACEAE

Cedrela trainii Arnold
(Plate 20, fig. 7)

Cedrela trainii Arnold, Amer. Midl. Naturalist 17, p. 1018, figs. 1, 2, 1936.
    Arnold, Contr. Univ. Mich., Mus. Pal. 5, no. 8, p. 95, pl. 6, figs. 1-3, 6, 1937.
    Chaney and Axelrod, Carnegie Inst. Wash. Pub. 617, p. 189, pl. 38, figs. 3, 5-9,
    1959 (see synonymy and discussion).

The record of this species in the flora is based on two specimens. One of these is a portion of a leaf with several fragmentary leaflets, the other represents part of a single leaflet. They display the typical camptodromous secondary venation of *Cedrela* leaflets, with the secondaries diverging at high angles and then looping upward at a low angle and disappearing along the margin, which they supply with a series of finer cross-ties. The tertiaries form an irregular polygonal network that encloses a similar pattern of finer 4th-order veinlets. The shape and venation of the leaflets appear essentially identical to that displayed by leaves of the modern *C. mexicana* Roemer as well as *C. odorata* Linnaeus of the West Indies. The winged fruits that are so typical of the genus have not been encountered in this flora, though they have been illustrated for the nearby Middlegate flora to the northwest (Axelrod, 1985).

Occurrence: Buffalo Canyon, hypotypes 9683-9684.

Family VACCINIACEAE

Vaccinium sophoroides (Knowlton) Brown
(Plate 21, fig. 3)

Vaccinium sophoroides (Knowlton) Brown, U.S. Geol. Surv. Prof. Paper 186J, p. 18, pl. 61, figs. 1-3, 11, 1937 (see synonymy).
   Chaney and Axelrod, Carnegie Inst. Wash. Pub. 617, p. 200, pl. 44, figs. 6-8, 1959.

A single leaf in the flora is similar to those in the Latah, Stinking Water, and Blue Mountains floras of Oregon and Washington. The leaf is obovate, 3.2 cm long, with a short petiole. The leaf is entire-margined, has a sharply acute tip, the secondaries are camptodrome, and the tertiaries form a coarse irregular polygonal network.

Leaves of the living *V. arboreum* (Marshall) Nuttall of the eastern United States are similar to the fossil, as are those of *V. parvifolium* Smith of the Pacific coastal strip from central California northward.

Occurrence: Buffalo Canyon, hypotype no. 9707.

Family ERICACEAE

Arbutus trainii Axelrod
(Plate 21, figs. 1, 2, 6, 7)

Arbutus trainii MacGinitie, Carnegie Inst. Wash. Pub. 416, p. 64, pl. 13, figs. 1, 2; pl. 12, fig. 3, 1933.
   Brooks, Carnegie Mus. Annals 24, p. 300, pl. 21, fig. 3, 1935.
Arbutus idahoensis (Knowlton) Brown. Brown, U.S. Geol. Surv. Prof. Paper 186-J, p. 184, pl. 59, figs. 2-4, 1937.
   Brooks, Amer. Midl. Naturalist 25, p. 519, pl. 14, fig. 4, 1941.
Arbutus prexalapensis Axelrod, Univ. Calif. Pub. Geol. Sci. 33, p. 310, pl. 32, figs. 1-3, 1956; ibid. 129, p. 182, pl. 13, fig. 13; pl. 15, fig. 1; pl. 34, fig. 10, 1985.

Gordonia idahoensis (Knowlton) Berry.   Chaney and Axelrod, Carnegie Inst. Wash.
       Pub. 617, p. 197, pl. 43, fig. 6, 1959.

    Arbutus leaves in the Buffalo Canyon flora are similar to those from the Trout Creek
(MacGinitie, 1933), Succor Creek (Brooks, 1935; Graham, 1963), Blue Mountains
(Chaney and Axelrod, 1959), and Fingerrock floras (Wolfe, 1964; UC Mus. Pal.). Most
of these were placed by Brown (1937) in a new combination, *A. idahoensis* (Knowlton)
Brown, based on the type Payette material (Knowlton, 1898).   Judging from the
illustration, the type Payette leaf is smaller and slenderer, as are those from the Hog
Creek (Dorf, 1936) and Stewart Spring (Wolfe, 1964) floras. Since these latter fossils
may represent a derived species, it seems best to keep them separate until supplementary
(unfigured) specimens in these floras can be compaed and the problem clarified.   The
size differences may only reflect local environment, much as noted for the samaras of
Acer oregonianum Knowlton discussed above.
    In this regard, the large suites of leaves from the Fingerrock and nearby Golddyke
florule, as well as those from the Middlegate (Axelrod, 1956) and Buffalo Canyon floras,
tend to be entire-margined, whereas crenate-serrate leaves are less frequent or rare as
compared with specimens in the floras of Oregon and Idaho.   Whether these represent
a varietal difference that reflects a somewhat warmer climate in the south is not presently
clear.
    The affinities of these *Arbutus* leaves is generally with members of the *A. xalapensis*
complex of Mexico, which also shows considerable variation over its range.
    Occurrence: Buffalo Canyon, hypotypes 9685-9687; homeotypes 9688-9692.

Family MYRTACEAE

Eugenia nevadensis Axelrod
(Plate 20, fig. 8)

Eugenia nevadensis Axelrod, Univ. Calif. Pub. Geol. Sci. 129, p. 180, pl. 34, fig. 9,
       1985.

    Several lanceolate leaves with numerous subparallel secondaries and a conspicuous
marginal vein represent this species.   The marginal vein is formed by secondaries that
divide just within the margin, then branch to join the adjacent secondary.   Similar,
though larger, leaves are recorded in the Florissant (MacGinitie, 1953, pl. 71) and Green
River floras (MacGinitie, 1969, pl. 18, 19), consistent with the warmer climate in which
they lived.   They represent different species but otherwise are similar to the present
material. MacGinitie compared these older species with leaves of *E. fluvatilis* Hemsley
and *E. jambos* Linnaeus.
    The species was rare in Nevada at this time. It occurs elsewhere in the Fingerrock
flora and also in the Carson Pass assemblage from the present Sierran crest at treeline.
    Occurrence: Buffalo Canyon, hypotype 9693, homeotypes 9694-9695.

Family OLEACEAE

Fraxinus desatoyana Axelrod, n. sp.
(Plate 20, figs. 1-6)

Description: Leaflets asymmetrically lanceolate, decurrent; tip acute to acuminate of the 2 complete specimens; one measures 7.5 cm long and 1.5 cm broad, the other 4.5 by 2.0 cm; secondaries widely spaced, diverging at high to moderate angles, camptodrome; tertiaries quite irregular, enclosing a finer irregular quaternary set of veins that enclose a finer irregular mesh with open endings; margin entire, somewhat undulate, or with occasional very small teeth; petiolule 4 mm long; texture medium.

Samaras slender, averaging about 2 cm long, with wing attached to middle of seed, the seed long-elliptic.

Discussion: These leaflets and samaras compare favorably with those produced by *F. velutina* Torrey. This small tree is confined to stream borders and ranges from the upper desert into the lower conifer forest at elevations near 2,000 m from southern California to western Texas and into northern Mexico.

This species differs from *F. caudata* Dorf (1930, pl. 13, fig. 8 only) in having more lanceolate, narrower leaflets. The samaras referred to *F. coulteri* Dorf (1936, pl. 3, figs. 3, 4) may represent *F. desatoyana*. Since samaras are not easy to separate specifically, it seems best to keep the taxa separate, at least temporarily, for *F. desatoyana* is more nearly like *F. velutina* than *F. oregona* Nuttall, which has been compared with *F. coulteri*. Samaras of *F. alcornii* Axelrod (1956, pl. 9, figs. 4, 5, 7-9) from the Horsethief Canyon locality of the Aldrich Station flora are slenderer and have also been compared with those of *F. velutina*; leaflets of ash were not found at that locality. In view of the more complete material at Buffalo Canyon, it seems best to keep these taxa separate until more material becomes available from the Aldrich Station area or other floras in the region.

Occurrence: Buffalo Canyon, holotype 9696; paratypes 9697-9699, 9702, 9703, 9706.

Fraxinus eastgatensis Axelrod, n. sp.
(Plate 22, figs. 4-7)

Description: Leaflet terminal, broadly ovate, 6.5 cm long including petiole, blade slightly decurrent on one side where it joins petiole; leaflet measures 3.7 cm broad in lower third of blade; widely spaced 4-5 secondaries, broadly looping up toward margin, camptodromous, giving off tertiaries to supply marginal blunt small teeth which are few and scattered; strong tertiaries depart from secondaries in lower half of blade, trend to marginal area and then divide into quaternaries that supply marginal teeth; intercostal area with tertiary cross-ties, forming a complex dense mesh, enclosing 4th- and 5th-order small veins that end in open polygonal mesh.

Samaras 2.0-2.8 cm long, seed about half as long, wing 0.6-0.7 cm broad, reaching to middle of seed, rounded above with strong midvein.

Discussion: This appears to be a new species of ash, one that differs considerably from those described previously from the region. *F. caudata* Dorf (1930) has much broader, larger leaflets with different venation; *F. alcornii* Axelrod (1956) has slenderer samaras; *F. dayana* Chaney and Axelrod (1959) has much broader samaras and is considerably larger. *F. desatoyana* has slenderer leaflets and samaras than *F. eastgatensis*. A general relationship is apparent with *F. chinensis* (=*F. hopeiensis*).

The specific name *eastgatensis* is derived from the old Pony Express station at Eastgate, situated 13 km northwest of Buffalo Canyon.

Occurrence: Buffalo Canyon, holotype 9895, paratypes 9705, 9701, 9700.

Appendix

Radiometric dates of dacitic tuffs in the Lower Member of the Buffalo Canyon Formation:

1. Dated by Krummenacher's Laboratory, San Diego State University, 1979. #1894. On Cuesta Ridge above corral directly west of fossil locality.
    Mineral: plagioclase. % K 3.14   % Ar (atmos.) 69.0

$$AGE = 18.03 + 1.1$$

2. Dated by University of Arizona, Isotope Geochem Laboratory, 1990. #UAKA 89 126. Welded tuff at "gate" in canyon east of flora, about 80 m. below flora.
    Mineral: feldspar concentrate. % K 4.96   % Ar (atmos.) 11.4

$$AGE = 17.5 + 0.4$$

3. Dated by Kruger Enterprises, Geochron Laboratory, 1990. #F-8788. Welded tuff along county road south of flora and about 90 m lower in section.
    Mineral: plagioclase. % K 7.55 average   % Ar. ppm 008519

$$AGE = 19.3 + 0.8$$

69

Literature Cited

AXELROD, D. I.

1937   A Pliocene flora from the Mount Eden beds, southern California. Carnegie Inst. Wash. Pub. 476:125-183.

1939   A Miocene flora from the western border of the Mohave Desert. Carnegie Inst. Wash. Pub. 516:1-128.

1944a   The Pliocene sequence in central California. Carnegie Inst. Wash. Pub. 553:207-224.

1944b   The Alvord Creek flora. Carnegie Inst. Wash. Pub. 553:225-262.

1956   Mio-Pliocene floras from west-central Nevada. Univ. Calif. Pub. Geol. Sci. 33:1-316.

1960   The evolution of flowering plants. *In* S. Tax, ed., Evolution After Darwin, vol. 1, The Evolution of Life, pp. 227-305. University of Chicago Press.

1962   A Pliocene *Sequoiadendron* forest from western Nevada. Univ. Calif. Pub. Geol. Sci. 39:195-268.

1964   The Miocene Trapper Creek flora of southern Idaho. Univ. Calif. Pub. Geol. Sci. 51:1-161.

1966   The Eocene Copper Basin flora of northeastern Nevada. Univ. Calif. Pub. Geol. Sci., 59:1-124.

1967   Geological history of the California insular flora. *In* R. N. Philbrick, ed., Proceedings of the Symposium on the Biology of the Channel Islands, pp. 267-316. Santa Barbara Botanical Garden.

1968   Tertiary floras and topographic history of the Snake River Basin, Idaho. Geol. Soc. Amer. Bull. 79:713-734.

1976a   History of the conifer forests, California and Nevada. Univ. Calif. Pub. Botany 70:1-62.

1976b   Evolution of the Santa Lucia fir (*Abies bracteata*) ecosystem. Ann. Missouri Bot. Gard. 63:24-41.

1977   Outline history of California vegetation. *In* M. G. Barbour and J. Major, eds., Terrestrial Vegetation of California, pp. 139-193. Wiley, New York.

1979   Desert vegetation, its age and origin. *In* J. R. Goodin and D. K. Northington, eds., Arid Land Plant Resources, pp. 1-72. International Center for Arid and Semiarid Land Studies, Texas Technical University, Lubbock.

1980a   Contributions to the Neogene Paleobotany of Central California. Univ. Calif. Pub. Geol. Sci. 121:1-212.

1980b   History of the maritime closed-cone pines, Alta and Baja California. Univ. Calif. Pub. Geol. Sci. 120:1-143.

1981a   Holocene climatic changes in relation to vegetation disjunction and speciation. Amer. Naturalist 117:847-870.

1981b   Altitudes of Tertiary forests estimated from paleotemperature. Proc. Sympos. on Qinghai-Xizang (Tibet) Plateau (Beijing, China). *In* Geological and Ecological Studies of Qinghai-Xizang Plateau, vol. 1, Geology, Geological History and Origin of Qinghai-Xizang Plateau, pp. 131-137. Science Press, Beijing and Gordon and Breach, Science Publishers, New York.

1982   Age and origin of the Monterey endemic area. Madroño 29:127-147.

1985   Miocene floras from the Middlegate basin, west-central Nevada. Univ. Calif. Pub. Geol. Sci. 129:1-277.

1987   The late Oligocene Creede flora, Colorado. Univ. Calif. Pub. Geol. Sci. 130:1-235.

1988   An interpretation of high montane conifers in western Tertiary floras. Paleobiology 14:301-306.

1990   The Middle Miocene Pyramid flora of western Nevada. MS. (For: Univ. Calif. Pub. Geol. Sci., 104 pp.)

——, and H. P. BAILEY

1976   Tertiary vegetation, climate, and altitude of the Rio Grande depression, New Mexico-Colorado. Paleobiology 2:235-254.

——, and P. H. RAVEN

1972   Evolutionary biogeography viewed from plate tectonic theory. *In* J. A. Behnke, ed., Challenging Biological Problems: Directions toward their solution, pp. 218-236. Oxford University Press. New York.

BAILEY, D. K.

1970   Phytogeography and taxonomy of *Pinus* subsection *Balfourianae*. Ann. Missouri Bot. Gard. 57:201-249.

BAILEY, H. P.

1960   A method of determining the warmth and temperateness of climate. Geografisker Annaler 42:1-16.

1964   Toward a unified concept of the temperate climate. Geogr. Review 54:516-545.

1966   The mean annual range and standard deviation as measures of dispersion of temperature around the annual mean. Geografisker Annaler 48A:183-194.

BARROWS, K. J.
    1971   Geology of the southern Desatoya Mountains, Churchill and Lander
            Counties, Nevada, Ph.D. Thesis. University of California, Los Angeles.
            349 pp.
BECKER, H. F.
    1961   Oligocene plants from the Upper Ruby River Basin, southwestern
            Montana. Geol. Soc. Amer. Mem. 82:1-127.
BERGER, R., and W. F. LIBBY
    1966   UCLA Radiocarbon dates, Part 5. Radiocarbon 8:467-497.
BERRY, E. W.
    1929   A revision of the flora of the Latah Formation. U.S. Geol. Surv. Prof.
            Paper 154-H:225-264.
    1931   A Miocene flora from Grand Coulee, Washington. U.S. Geol. Surv.
            Prof. Paper 170-C:31-42.
BROWN, R. W.
    1937   Additions to some fossil floras of the western United States. U.S. Geol.
            Surv. Prof. Paper 186-J:163-206.
BURKE, D. B., and E. H. McKEE
    1979   Mid-Cenozoic volcano-tectonic troughs in central Nevada. Geol. Soc.
            America Bull. 90:181-184.
CHANEY, R. W.
    1920   The flora of the Eagle Creek Formation. Contr. Walker Mus.,   2:115-
            181.
    1927   Geology and paleontology of the Crooked River basin, with special
            reference to the Bridge Creek flora. Carnegie Inst. Wash. Pub. 346:45-
            138.
    1944   The Dalles (Oregon) flora. Carnegie Inst. Wash. Pub. 553:385-321.
    1959   Miocene floras of the Columbia Plateau, Part 1. Composition and
            interpretation. Carnegie Inst. Wash. Pub. 617:1-134.
———, and D. I. AXELROD
    1959   Miocene floras of the Columbia Plateau, Part 2.   Systematic
            considerations. Carnegie Inst. Wash. Pub. 617:135-229.
CONDIT, C.
    1938   The San Pablo flora of west-central California. Carnegie Inst. Wash.
            Pub. 476:217-268.
    1944   The Table Mountain flora. Carnegie Inst. Wash. Pub. 553:57-90.
DALRYMPLE, G. B.
    1979   Critical tables for conversion of K-Ar ages from old to new constants.
            Geology 7:558-560.
DORF, E.
    1930   Pliocene floras of California. Carnegie Inst. Wash. Pub. 412:1-112.
    1936   A Late Tertiary flora from southwestern Idaho. Carnegie Inst. Wash.
            Pub. 476:73-124.

GRAHAM, A.
    1963   Systematic revision of the Sucker Creek and Trout Creek Miocene floras
            of southeastern Oregon. Amer. Jour. Bot. 50:921-936.
    1965   The Sucker Creek and Trout Creek Miocene floras of southeastern
            Oregon. Kent State Univ. Bull., Research Series 9:1-147.
KNOWLTON, F. H.
    1926   Flora of the Latah Formation of Spokane, Washington, and Coeur
            d'Alene, Idaho. U.S. Geol. Surv. Prof. Paper 140:17-81.
KORNAS, J.
    1972   Corresponding taxa and their ecological background in the forests of
            temperate Eurasia and North America. *In* D. H. Valentine, ed.,
            Taxonomy, Phytogeography and Evolution, pp. 37-60. Academic Press,
            London and New York.
LaMOTTE, R. A.
    1936   The Upper Cedarville flora of northwestern Nevada and adjacent
            California. Carnegie Inst. Wash. Pub. 455:57-142.
MacGINITIE, H. D.
    1933   The Trout Creek flora of southwestern Oregon. Carnegie Inst. Wash.
            Pub. 416:21-68.
    1953   Fossil plants of the Florissant beds, Colorado. Carnegie Inst. Wash.
            Pub. 599:1-188.
    1962   The Kilgore flora, a Late Miocene flora from northern Nebraska. Univ.
            Calif. Pub. Geol. Sci. 35:67-158.

    1969   The Eocene Green River flora of northwestern Colorado and
            northeastern Utah. Univ. Calif. Pub. Geol. Sci. 83.
McKEE, E. H.
    1974   Northumberland caldera and Northumberland tuff. Nevada Bur. Mines
            and Geol. Report 19:35-41.
MASON, H. L.
    1934   The Pleistocene Tomales flora. Carnegie Inst. Wash. Pub. 415:81-179.
    1947   Evolution of certain floristic associations in western North America.
            Ecol. Monogr. 17:159-183.
MUNZ, P. A., and D. D. KECK
    1959   A California Flora. University of California Press, 1681 pp. Berkeley,
            Los Angeles.
NEWBERRY, J. S.
    1898   The later extinct floras of North America. U.S. Geol. Surv. Monog.
            35:1-295.
OLIVER, E.
    1934   A Miocene flora from the Blue Mountains, Oregon. Carnegie Inst.
            Wash. Pub. 455:1-27.
PECK, D. L., A. B. GRIGGS, H. G. SCHLICHER, F. G. WELLS, and H. M. DOLE
    1964   Geology of the central and northern parts of the western Cascade Range
            in Oregon. U.S. Geol. Surv. Prof. Paper 449:1-56.

REED, R. D.
  1933   Geology of California. Amer. Assoc. Petrol. Geol., Tulsa, Oklahoma. 355 pp. (Reprinted 1951.)
RENNEY, K. M.
  1972   The Miocene Temblor flora of west-central California. MA thesis, University of California, Davis. 106 pp.
SAWYER, J. O., and D. A. THORNBURGH
  1977   Montane and subalpine vegetation of the Klamath Mountains. *In* M. G. Barbour and J. Major, eds., Terrestrial Vegetation of California, p. 699-732, Wiley, New York.
SMILEY, C. J.
  1963   The Ellensburg flora of Washington. Univ. Calif. Pub. Geol. Sci. 35:159-276.
——, and W. C. REMBER
  1985   Composition of the Miocene Clarkia flora. *In* C. J. Smiley, ed., Cenozoic History of the Pacific Northwest: Interdisciplinary studies on the Clarkia fossil beds of northern Idaho, pp. 95-112. Pacific Division, American Association for the Advancement of Science.
SMITH, H. V.
  1938   Notes on fossil plants from Hog Creek in southwestern Idaho. Papers Mich. Acad. Sci. Arts and Letters 23:223-231.
  1939a   Additions to the fossil flora of Sucker Creek, Oregon. Papers Mich. Acad. Sci. Arts and Letters 24:107-120.
  1941   A Miocene flora from Thorn Creek, Idaho. Amer. Mid. Nat. 25:473-522.
SPICER, R. A., and J. A. WOLFE.
  1987   Plant taphonomy of Holocene deposits in Trinity (Clair Engle) Lake, northern California. Paleobiology 13:227-245.
STEBBINS, G. L., and J. MAJOR
  1965   Endemism and speciation in the California flora. Ecol. Monogr. 35:1-35.
TANAI, T.
  1970   The Oligocene floras from the Kushiro coal fields, Hokkaido, Japan. Jour. Faculty Sci. (Hokkaido University) 14:383-514.
——, and J. WOLFE
  1977   Revisions of *Ulmus* and *Zelkova* in the middle and late Tertiary of western North America. U.S. Geol. Surv. Prof. Paper 1026:1-14.
TURNER, D. L.
  1970   Potassium-argon dating of Pacific Coast Miocene foraminiferal stages. Geol. Soc. Amer. Spec. Paper 124:91-129.
WHITTAKER, R. H.
  1960   Vegetation of the Siskiyou Mountains, Oregon and California. Ecol. Monogr. 30:279-338.
WILLDEN, R., and R. S. SPEED
  1974   Geology and mineral deposits of Churchill County, Nevada. Nevada Bur. Mines and Geol. Bull. 83:1-95.

WOLFE, J. A.
    1964a   Miocene floras from Fingerrock Wash, southwestern Nevada. U.S. Geol. Surv. Prof. Paper 454-N:1-36.

    1964b   Neogene floristic and vegetational history of the Pacific Northwest. Madroño 20:83-110.

    1966   Tertiary plants from the Cook Inlet region, Alaska. U.S. Geol. Surv. Prof. Paper 398-B:1-32.

    1972   An interpretation of Alaskan Tertiary floras. *In*: A. Graham, ed., Floristics and Paleofloristics of Asia and Eastern North America, pp. 201-233, Elsevier, Amsterdam.

    1977   Paleogene floras from the Gulf of Alaska region. U.S. Geol. Surv. Prof. Paper 997:1-108.

    1980   The Miocene Seldovia Point flora from the Kenai Group, Alaska. U.S. Geol. Surv. Prof. Paper 1105.

——, and T. TANAI
    1987   Systematics, phylogeny, and distribution of *Acer* (maples) in the Cenozoic of western North America. Jour. Fac. Sci. (Hokkaido University) 22:1-246.

Plates

PLATE 1

Fossil Locality and Allied Modern Vegetation

Fig. 1. The Buffalo Canyon flora is on the south slope across the valley where the fossil beds dip gently west.

Fig. 2. Modern vegetation in the Klamath Mountains of northwestern California has many conifer and sclerophyll species allied to those in the fossil flora.

PLATE 2

Modern Vegetation Allied to the Fossil Flora

Fig. 1. Conifer-deciduous hardwood forest in the Adirondack Mountains has species allied to those in the Buffalo Canyon flora.

Fig. 2. Modern conifer-deciduous hardwood forest in Porcupine Mountains of Michigan has several species allied to those in the fossil flora.

Plate 3

Fig. 1   *Abies concoloroides* Brown.   Hypotype 7920.
Fig. 2.   *Abies laticarpus* MacGinitie.   Hypotype 7922.
Figs. 3-6.   *Pinus balfouroides* Axelrod.   Hypotypes 7983, 7981, 7984, 7986.
Figs. 7-10.   *Picea lahontense* MacGinitie.   Hypotypes 7956, 7954, 7955, 9347.
Figs. 11-12.   *Pinus ponderosoides* Axelrod.   Hypotypes 7993, 7991.
Figs. 13-16.   *Picea sonomensis* Axelrod.   Hypotypes 7968, 7966, 7969, 7967.
Figs. 17-19.   *Pinus balfouroides* Axelrod.   Hypotypes 7998, 7997, 7999.

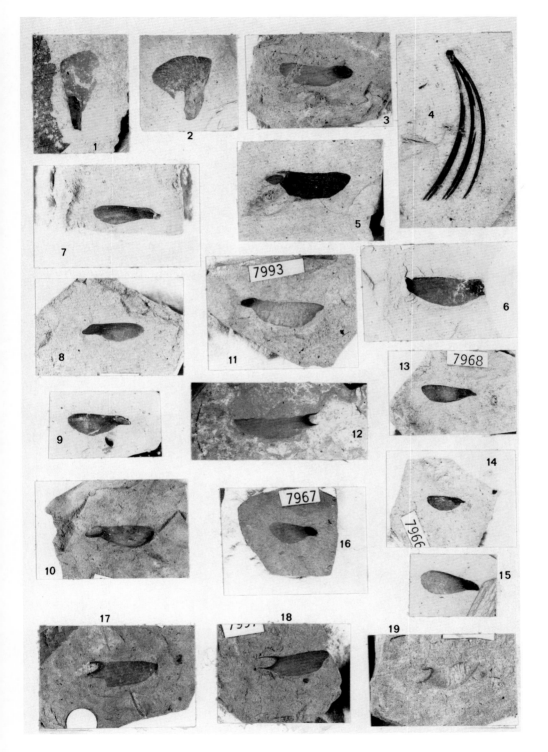

Plate 4

Figs. 1-5.   *Pseudotsuga sonomensis* Dorf. Hypotypes 9324, 9320, 9321, 9322, 9323.
Figs. 6-9.   *Tsuga mertensioides* Axelrod. Hypotypes 9335, 9333, 9337, 9334.
Figs. 10-13.   *Picea magna* MacGinitie. Hypotypes 7945, 7943, 7948, 7942.
Figs. 14-17.   *Juniperus desatoyana* Axelrod. Holotype 9355, paratypes 9351 (fruit), 9348, 9350.
Fig. 18.   *Chamaecyparis cordillerae* Edwards and Schorn. Hypotype 9892.

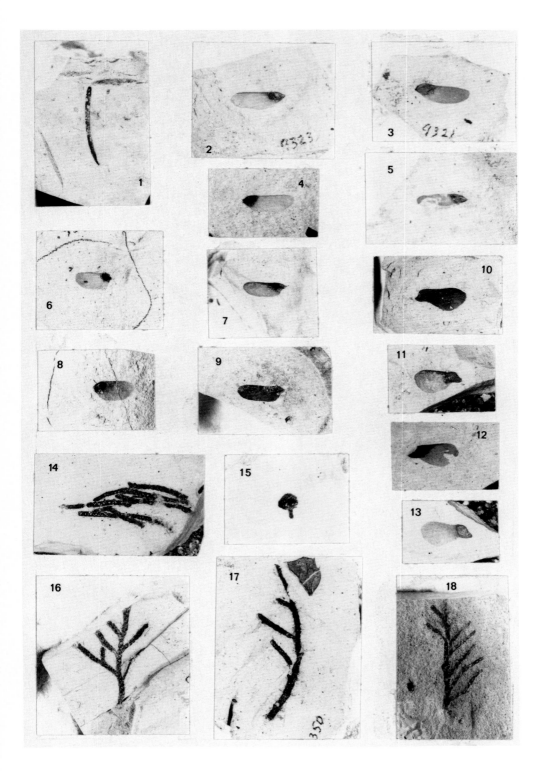

BUFFALO CANYON FOSSILS

Plate 5

Fig. 1.   *Typha lesquereuxi* Cockerell.   Hypotype 9358.
Figs. 2-5.   *Larix churchillensis* Axelrod.   Hypotypes 7937, 7939, 7938, 7936.
Fig. 6   *Populus pliotremuloides* Axelrod.   Hypotype 9384.
Figs. 7-11.   *Populus cedrusensis* Wolfe.   Hypotypes 9367, 9368, 9366, 9377, 9369.

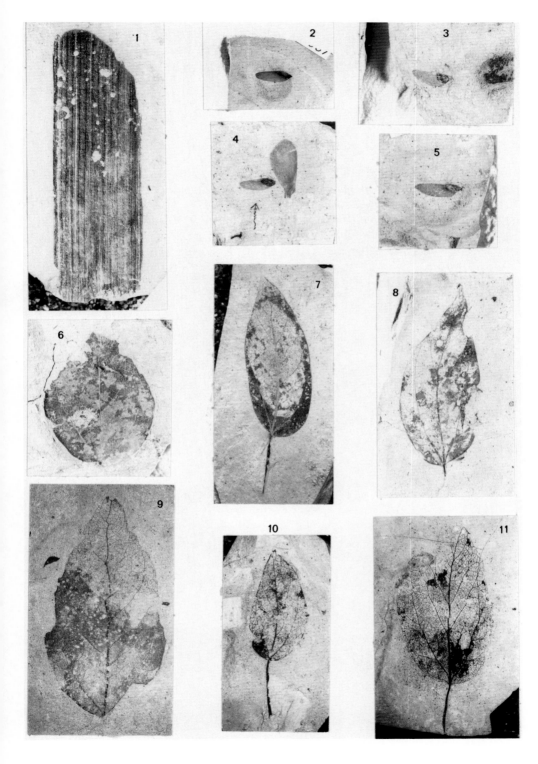

Plate 6

Figs. 1, 3.   *Populus eotremuloides* Knowlton. Hypotypes 9378, 9380.
Fig. 2.   *Salix storeyana* Axelrod. Hypotype 9432.
Figs. 4-7.   *Salix churchillensis* Axelrod. Holotype 9390, paratypes 9389, 9391, 9392.
Fig. 8.   *Populus eotremuloides* Knowlton. Hypotype 9379.

BUFFALO CANYON FOSSILS

Plate 7

Figs. 1-4, 6.   *Salix storeyana* Axelrod. Hypotypes 9429, 9431, 9430, 9439, 9435.
Fig. 5.   *Salix pelviga* Wolfe. Hypotype 9435.
Figs. 7, 8.   *Salix* sp. capsules. Nos. 9441, 9440.
Figs. 9-11.   *Populus pliotremuloides* Axelrod. Hypotypes 9386, 9388, 9382.

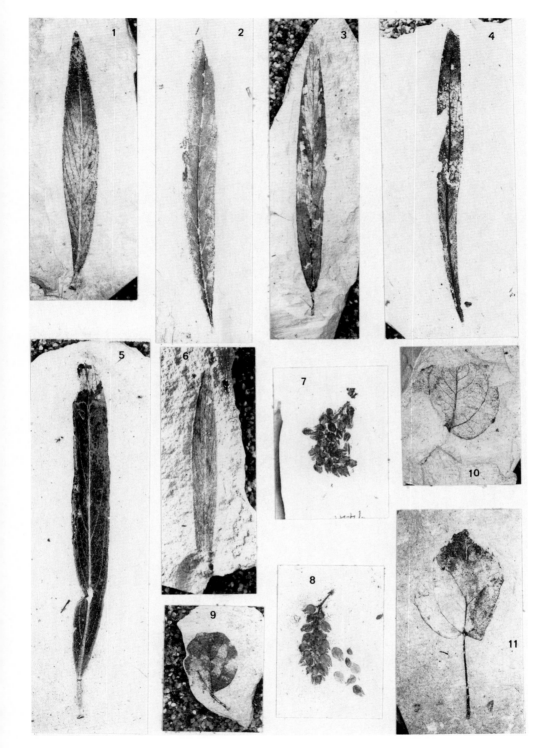

Plate 8

Figs. 1, 2. *Salix owyheeana* Chaney and Axelrod. Hypotypes 9414, 9415.
Figs. 3, 4. *Salix laevigatoides* Axelrod. Hypotypes 9408, 9409.
Figs. 5, 6. *Salix pelviga* Wolfe. Hypotypes 9419, 9420.

BUFFALO CANYON FOSSILS

Plate 9

Figs. 1-3.   *Betula thor* Knowlton. Hypotypes 9485, 9449, 9486.
Fig. 4.   *Betula idahoensis* Smith. Hypotype 9484.
Figs. 5, 7.   *Betula thor* Knowlton. Hypotypes 9488, 9490.
Fig. 6.   *Carpinus oregonensis* Axelrod. Hypotype 9497.

BUFFALO CANYON FOSSILS

Plate 10

Figs. 1-5.   *Betula desatoyana* Axelrod.   Holotype 9476, paratypes 9470, 9472, 9469, 9471.

Figs. 6-11.   *Betula ashleyi* Axelrod.   Hypotypes 9462, 9448, 9463, 9468, 9451, 9449.

BUFFALO CANYON FOSSILS

Plate 11

Figs. 1, 2. *Alnus latahensis* Axelrod. Hypotypes 9446, 9445.
Figs. 3, 4. *Alnus* sp., cones of. Nos. 9444, 9447.
Figs. 5-7. *Carya bendirei* (Lesq.) Chaney and Axelrod. Hypotypes 9498, 9499, 9500.

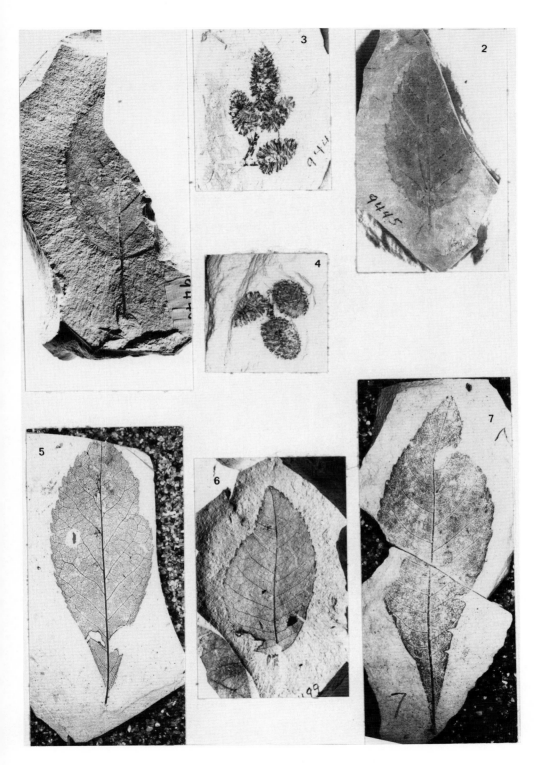

BUFFALO CANYON FOSSILS

Plate 12

Figs. 1-3.   *Quercus hannibalii* Dorf. Hypotypes 9519, 9524, 9523.
Figs. 4, 5.   *Quercus hannibalii* Dorf. Acorn and cup. Hypotypes 9520, 9521.
Figs. 6, 7.   *Quercus hannibalii* Dorf. Hypotypes 9546, 9547.
Figs. 8, 9.   *Ulmus speciosa* Newberry. Samaras. Hypotypes 9555, 9554.
Fig. 10.   *Ulmus speciosa* Newberry. Hypotype 9553.
Figs. 11, 12.   *Quercus wislizenoides* Axelrod. Hypotypes 9548, 9549.

Plate 13

Figs. 1-4.   *Juglans desatoyana* Axelrod. Holotype 9507; paratypes 9511, 9508, 9509.

Figs. 5-9.   *Zelkova brownii* Tanai and Wolfe. Hypotypes 9572, 9568, 9566, 9565, 9567.

Fig. 10.   *Ulmus speciosa* Newberry. Hypotype 9557.

Plate 14

Figs. 1, 2.   *Chrysolepis sonomensis* Axelrod. Hypotypes 9540, 9541.
Figs. 3, 4.   *Mahonia macginitieii* Axelrod. Hypotypes 9585, 9584.
Figs. 5, 6, 8.   *Mahonia reticulata* (MacGinitie) Brown. Hypotypes 9590, 9588, 9587.
Figs. 7, 9.   *Mahonia macginitieii* Axelrod. Hypotypes 9583, 9582.

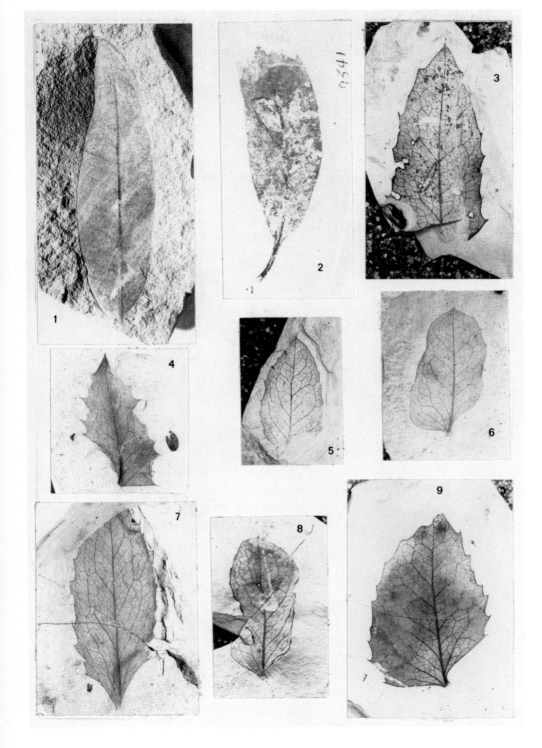

BUFFALO CANYON FOSSILS

Plate 15

Fig. 1.   *Ribes bonhamii* Axelrod. Holotype 9608.
Figs. 2-4.   *Ribes barrowsii* Axelrod. Holotype 9604, paratypes 9605, 9606.
Figs. 5, 6.   *Ribes webbii* Wolfe. Hypotypes 9611, 9613.
Figs. 7-9.   *Hydrangea bendirei* (Ward) Knowlton. Hypotypes 9601, 9603, 9602.
Fig. 10.   *Ribes stanfordianum* Dorf. Hypotype 9610.

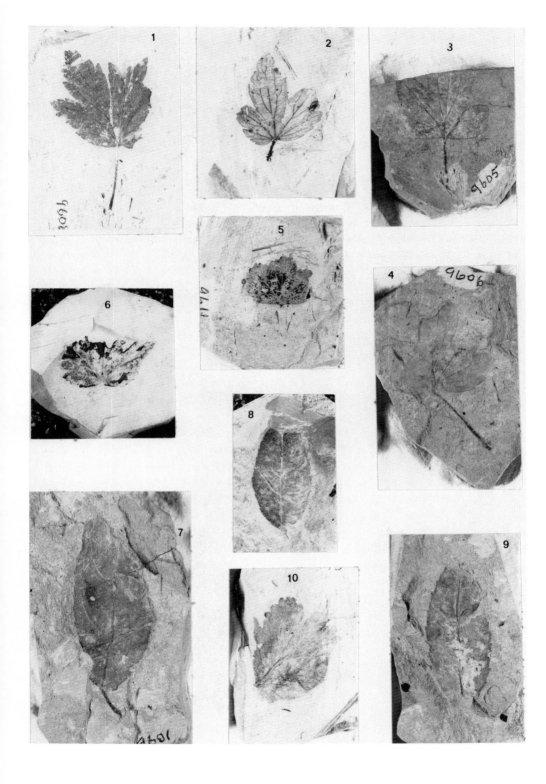

BUFFALO CANYON FOSSILS

Plate 16

Figs. 1-7.   *Amelanchier desatoyana* Axelrod.   Holotype 9614, paratypes 9615,
             9617, 9619, 9618, 9616, 9620.
Fig. 8.   *Cercocarpus ovatifolius* Axelrod. Hypotype 9621.
Figs. 9, 10. *Crataegus middlegatei* Axelrod. Hypotypes 9622, 9623.
Fig. 11.   *Heteromeles desatoyana* Axelrod. Holotype 9625.

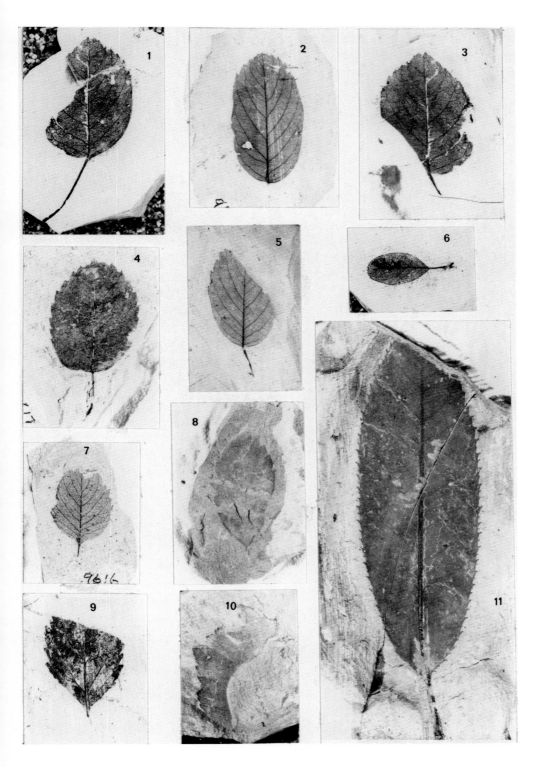

BUFFALO CANYON FOSSILS

Plate 17

Figs. 1, 3.   *Prunus chaneyii* Condit. Hypotypes 9635, 9636.
Fig. 2.   *Prunus moragensis* Axelrod. Hypotype 9639.
Fig. 4.   *Chamaebatia nevadensis* Axelrod. Holotype 9710.
Figs. 5, 6.   *Nymphaeites nevadensis* (Knowlton) Brown. Hypotypes 9596, 9595.
Figs. 7-9.   *Prunus treasheri* Chaney. Hypotypes 9643, 9642, 9640.

BUFFALO CANYON FOSSILS

Plate 18

Figs. 1, 2.   *Lyonothamnus parvifolius* (Axelrod) Wolfe. Hypotypes 9629, 9630.
Figs. 3-5.   *Rosa harneyana* Chaney and Axelrod. Hypotypes 9647, 9645, 9648.
Figs. 6-8.   *Sorbus cassiana* Axelrod. Hypotypes 9650, 9652, 9651.

Plate 19

Figs. 1-3.   *Acer oregonianum* Knowlton. Hypotypes 9679, 9678, 9893.
Fig. 4.   *Acer negundoides* MacGinitie. Hypotype 9894.
Figs. 5, 6. *Amorpha stenophylla* Axelrod. Holotype 9660, paratype 9661.
Fig. 7.   *Acer tyrrellii* Smiley. Hypotype 9682.
Fig. 8.   *Acer trainii* Wolfe and Tanai. Hypotype 9677,
Figs. 9, 10. *Acer negundoides* MacGinitie. Hypotypes 9667, 9668.
Fig. 11.   *Acer medianum* Knowlton. Hypotype 9664.
Fig. 12.   *Robinia bisonensis* Axelrod. Holotype 9662.
Fig. 13.   *Acer negundoides* MacGinitie. Hypotype 9666.

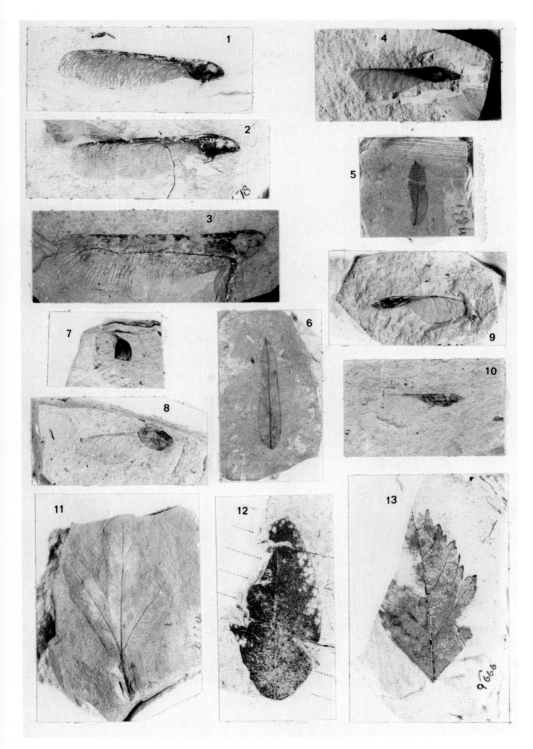

BUFFALO CANYON FOSSILS

Plate 20

Figs. 1-3.   *Fraxinus desatoyana* Axelrod. Holotype 9696, paratypes 9697, 9698.
Figs. 4-6.   *Fraxinus desatoyana* Axelrod. Paratypes 9699, 9706, 9703.
Fig. 7.   *Cedrela trainii* Arnold. Hypotype 9683.
Fig. 8.   *Eugenia nevadensis* Axelrod. Hypotype 9693.

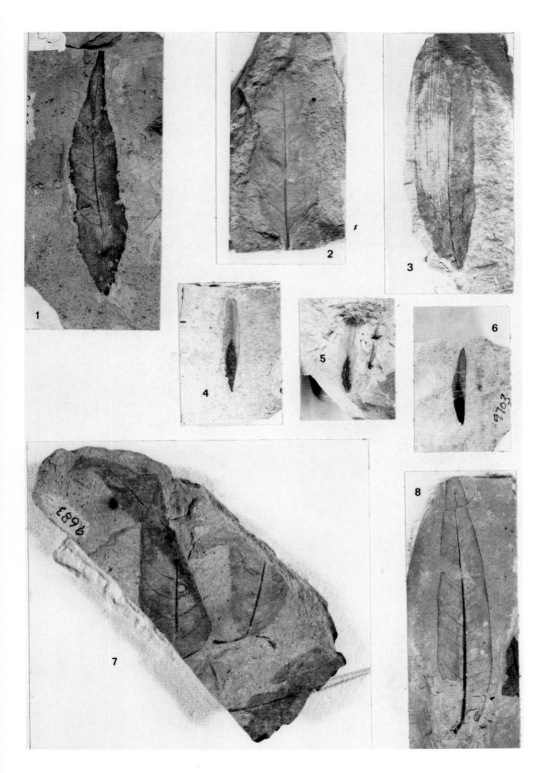

BUFFALO CANYON FOSSILS

Plate 21

Figs. 1, 2.   *Arbutus prexalapensis* Axelrod. Hypotypes 9690, 9692.
Fig. 3.   *Vaccinium sophoroides* (Knowlton) Brown. Hypotype 9707.
Figs. 4, 5.   *Symphoricarpos wassukana* Axelrod. Hypotypes 9708, 9709.
Figs. 6, 7.   *Arbutus prexalapensis* Axelrod. Hypotypes 9687, 9685.

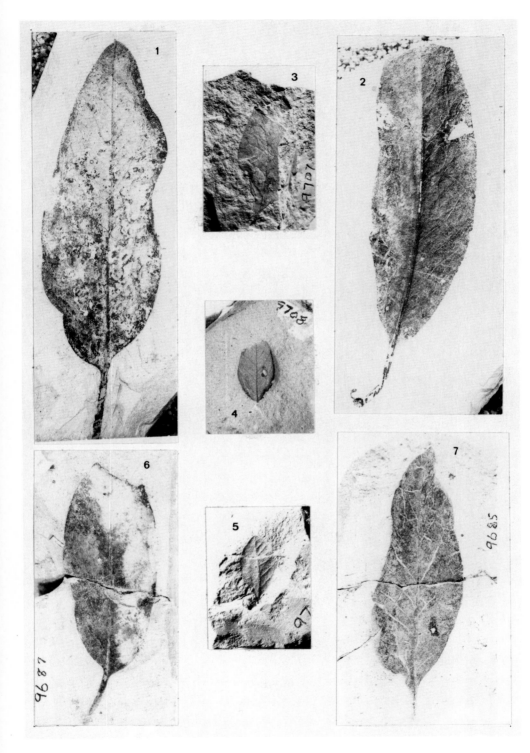

BUFFALO CANYON FOSSILS

Plate 22

Figs. 1-3.   *Populus payettensis* (Knowlton) Axelrod. Hypotypes 9888, 9886, 9887.
Figs. 4-7.   *Fraxinus eastgatensis* Axelrod. Holotype 9895, paratypes 9705, 9701, 9700.
Fig. 8.   *Salix desatoyana* Axelrod. Holotype 6881.